Vidyadhar S. Mandrekar

Weak Convergence of Stochastic Processes

De Gruyter Graduate

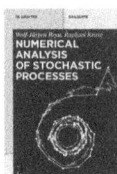

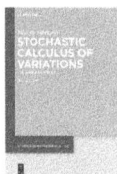

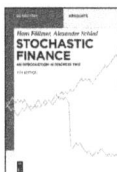

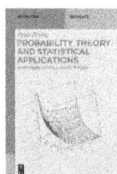

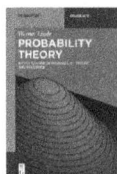

Vidyadhar S. Mandrekar

Weak Convergence of Stochastic Processes

With Applications to Statistical Limit Theorems

DE GRUYTER

Mathematics Subject Classification 2010
60B20, 60F17

Author
Vidyadhar S. Mandrekar
Michigan State University
Department of Statistics and Probability
East Lansing MI 48824
USA
atma1m@gmail.com

ISBN 978-3-11-047542-5
e-ISBN (PDF) 978-3-11-047631-6
e-ISBN (EPUB) 978-3-11-047545-6

Library of Congress Cataloging-in-Publication Data
A CIP catalog record for this book has been applied for at the Library of Congress.

Bibliographic information published by the Deutsche Nationalbibliothek
The Deutsche Nationalbibliothek lists this publication in the Deutsche Nationalbibliografie; detailed bibliographic data are available on the Internet at http://dnb.dnb.de.

Typesetting: Compuscript Ltd, Ireland
Printing and binding: CPI books GmbH, Leck
Cover image: 123dartist/iStock/thinkstock
♾ Printed on acid-free paper
Printed in Germany

www.degruyter.com

Contents

1 Weak convergence of stochastic processes

Introduction

The study of limit theorems in probability has been important in inference in statistics. The classical limit theorems involved methods of characteristic functions, in other words, Fourier transform. One can see this in the recent book of Durrett [8]. To study the weak convergence of stochastic processes, one may be tempted to create concepts of Fourier's theory in infinite dimensional cases. However, such an attempt fails, as can be seen from the work of Sazanov and Gross [14]. The problem was also solved by Donsker [5] by choosing techniques of weak convergence of measures on the space of continuous functions by interpolating the partial sum in the central limit theorem. If we denote for $X_1, \ldots, X_n$ i.i.d. (independent identically distributed random variable with zero expectation and finite variance)

$$S_{n,m} = X_1 + \ldots + X_{m/\sqrt{n}}, \text{ where } m = 1, 2, \ldots, n,$$

and $S_{n,0} = 0$. One can define

$$S_{n,u} = \begin{cases} S_{n,m} \text{ if } u = 0, 1, \ldots n \\ \text{linear if } u \epsilon [m-1, m] \end{cases}$$

Then, the continuous process $S_{[n,t]}$, $t \epsilon [0, 1]$ generates a sequence of measures on $C[0, 1]$ if we can prove, with appropriate definition P_n converges weakly to the measure P given by Brownian motion on $C[0, 1]$, we can conclude that

$$\max_{0 \le t \le 1} |S_{n,t}| \text{ converges to } \max_{0 \le t \le 1} |W(t)|$$

and

$$\frac{R_n}{\sqrt{n}} = \frac{1}{\sqrt{n}} + \max_{1 \le m \le n} S_{n,m} - \min_{0 \le t \le 1} S_{n,m}$$

converges to

$$\max_{1 \le t \le 1} W(t) - \min_{0 \le t \le 1} W(t),$$

giving more information than the classical central theorem. One can also prove empirical distribution function, if properly interpolated,

$$\hat{F}_n(x) = \frac{1}{n} \sum_{1}^{n} 1(x_i \le x), x \epsilon \mathbb{R},$$

satisfies

$$\sqrt{n} \sup_x |\hat{F}_n(x) - F(x)|,$$

which converges to the maximum of the Brownian bridge $\max_{0 \le t \le 1} |W(t) - W(1)|$, justifying the Kolmogorov-Smirnov statistics for F continuous distribution function of X_1. We prove these results using convergence in probability in $C[0, 1]$ space of $S[n.]$ to $W[.]$ in the supremum distance in $C[0, 1]$. For this, we used the embedding theorem of Skorokhod [23]. We follow [8] for the proof of this theorem of Skorokhod.

As we are interested in the convergence of semi-martingales to prove the statistical limit theorems for censored data that arise in clinical trials, we introduce the following Billingsley [2] convergence in a separable metric (henceforth, Polish space). Here we show the compactness of the sequences of probability measures on Polish space using the so-called tightness condition of Prokhorov [22], which connects compactness of measures with a compact set having large measures for the whole sequence of probability measures. We then consider the form of compact sets in $C[0, 1]$ using the Arzela-Ascoli theorem. We also consider the question if one can define on the space of functions with jump discontinuity $D[0, T]$ a metric to make it a Polish space. Here we follow Billingsley [2] to present the so-called Skorokhod topology on $D[0, T]$ and $D[0, \infty)$. We then characterize the compact sets in these space to study the tightness of a sequence of probability measures. Then we use a remarkable result of Aldous [1] to consider compactness in terms of stopping times. This result is then exploited to study the weak convergence of semi-martingales in $D[0, \infty)$.

Following Durrett and Resnick [9], we then generalize the Skorokhod embedding theorem for sums of dependent random variables. This allows us to extend weak convergence result, as in Chapter 2, to martingale differences [3]. Using the work of Gordin [13], one can reduce a similar theorem for stationary sequences to that for martingale differences. If one observes that empirical measures take values in $D[0, 1]$ and we try to use the maximum norm, we get a nonseparable space. Thus, we can ask the question, “can one study weak convergence of stochastic processes taking values in non-separable metric spaces? This creates a problem with measurability as one can see from Dudley-Phillip’s [7] work presented in [18]. To handle these problems, we study the so-called empirical processes following Van der Vaart and Wellner [25]. We introduce covering numbers, symmetrization, and sub-Gaussian inequalities to find conditions for weak convergence of measures.

We begin chapter 2 by constructing a measure on $\mathbb{R}^T$ to get a stochastic process. We then construct a Gaussian process with given covariance. We obtain sufficient conditions on the moments of a process indexed by $T = \mathbb{R}$ to have a continuous version. Using this, we construct Brownian motion. Then we prove the Skorokhod embedding theorem for sums of independent random variables. It is then exploited to obtain a convergence in $C[0, 1]$ of the functions of the sum of independent random variables as described earlier. As a consequence of the central limit theorem in $C[0, 1]$, we prove using [20] the results of [11] and the convergence of symmetric statistics. The chapter ends by giving a weak convergence of probability measures and Prokhorov's tightness result on a Polish space. In chapter 3, we study compact sets in $C[0, 1]$ and use them to get alternate proof of theorems similar to the one in chapter 2 using weak convergence in Polish spaces. This is followed by studying the topology of Skorokhod on $D[0, T]$ and $D[0, \infty)$ and proving that these are Polish spaces. Again, we study the compact sets and prove the result of Aldous. Chapter 4 studies the weak convergence of semi-martingales, which requires Lenglart [16] inequality to prove compactness using the result of Aldous. As a consequence of this, we derive weak convergence of Nelson and Kaplan-Meier estimates by simplifying the proof of Gill [12]. We do not present convergence of the Susarla-Van Ryzin Bayes estimate [24], but it can be obtained by similar methods as shown in [4]. For convergence of Linden-Bell estimates arising in astronomy, see [21] where similar techniques are used. Chapter 5 considers limit theorems as in chapter 2 using generalization of Skorokhod theorem from [9]. Limit theorems in chapters 2 and 5 use techniques given in Durrett's book [8]. Chapter 3 follows the presentation in the book of Billingsley [2], and chapter 4 uses the simplified version of techniques in [17] (see also [15]). We present in the last chapter the convergence of empirical processes using the techniques mentioned above taken from [25].

Acknowledgments

The first version of this presentation was typed by Mr. J. Kim. Professor Jayant Deshpande encouraged me to write it as a book to make it available to wider audiences. Professor U. Naik-Nimbalkar read the final version. I thank all of them for their help in preparing the text. Finally, I thank Dr. K. Kieling for suggesting the format presented here.

2 Weak convergence in metric spaces

We begin in this chapter the process of associating a probability measure on the function space $\mathbb{R}^T$ for any set T given a family of probability measures on $\mathbb{R}^S$, $S \subseteq T$ finite set with certain conditions. This allows us to define a family of real random variables $\{X_t, t\epsilon T\}$ on a probability space $(\Omega, \mathcal{F}, P)$. Such a family is referred to as a stochastic process. This idea originated from the work of Wiener (see [19]) in the special case of Brownian motion. Kolmogorov generalized it for the construction of any stochastic process and gave conditions under which one can find a continuous version (cf. section 2.5), that is, a stochastic process with continuous paths. We use his approach to construct Brownian motion. We explain it in the next section.

2.1 Cylindrical measures

Let $\{X_t, t \in T\}$ be family of random variable on a probability space $(\Omega, \mathcal{F}, P)$. Assume that X_t takes values in $(\mathcal{X}_t, \mathcal{A}_t)$. For any finite set $S \subset T$,

$$\mathcal{X}_S = \prod_{t\in S} \mathcal{X}_t,\ \mathcal{A}_S = \bigotimes_{t\in S} \mathcal{A}_t,\ Q_S = P \circ (X_t, t \in S)^{-1},$$

where Q_S is the induced measure. Check, if $\Pi_{S'S} : \mathcal{X}_{S'} \to \mathcal{X}_S$ for $S \subset S'$, then

$$Q_S = Q_{S'} \circ \Pi_{S'S}^{-1} \tag{2.1}$$

Suppose we are given a family $\{Q_S, S \subseteq T \text{ finite-dimensional}\}$, a probability measure where Q_S on $(\mathcal{X}_S, \mathcal{A}_S)$. Assume they satisfy (2.1). Then, there exists Q on $(\mathcal{X}_T, \mathcal{A}_T)$ such that

$$Q \circ \Pi_S^{-1} = Q_S,$$

where $\mathcal{X}_T = \prod_{t\in T} \mathcal{X}_t$, $\mathcal{A}_T = \sigma\left(\bigcup_{S\subset T} \mathcal{C}_S\right)$, $\mathcal{C}_S = \Pi_S^{-1}(\mathcal{A}_S)$.

Remark: For

$$\begin{aligned} S \subset T&, \text{ finite}, C \in \mathcal{C}_S \\ \text{define } Q_0(C) &= Q_S(A), \text{ where } C = \Pi_S^{-1}(A) \\ \text{as } \mathcal{C}_S &= \Pi_S^{-1}(\mathcal{A}_S) \end{aligned}$$

We can define Q_0 on $\bigcup_{S\subset T} \mathcal{C}_S$. Then, for $C \in \mathcal{C}_S$ and $\mathcal{C}_{S'}$, then by (2.1),

$$Q_0(C) = Q_S(A) = Q_{S'}(A),$$

and hence Q_0 is well-defined.

Note that as

$$\mathcal{C}_{S_1} \cup \mathcal{C}_{S_2} \cup \cdots \cup \mathcal{C}_{S_k} \subset \mathcal{C}_{S_1 \cup \cdots \cup S_k}.$$

Q_0 is finitely additive on $\bigcup_{S \subset T} \mathcal{C}_S$. We have to show the countable additivity.

Definition 2.1: A collection of subsets $\mathcal{K} \subset \mathcal{X}$ is called a compact class if for every sequence $\{C_k, k = 1, 2, \cdots n\}$, n finite,

$$\bigcap_{k=1}^{n} C_k \neq \emptyset \Longrightarrow \bigcap_{k=1}^{\infty} C_k \neq \emptyset$$

Exercise 1: Every subcollection of compact class is compact.

Exercise 2: If $\mathcal{X}$ and $\mathcal{Y}$ are two spaces and $T : \mathcal{X} \to \mathcal{Y}$ and $\mathcal{K}$ is a compact class in $\mathcal{Y}$, then $T^{-1}(\mathcal{K})$ is a compact class in $\mathcal{X}$.

Definition 2.2: A finitely additive measure μ on $(\mathcal{X}, \mathcal{A}_0)$ is called compact if there exists a compact class $\mathcal{K}$ such that for every $A \in \mathcal{A}_0$ and $\epsilon > 0$, there exists $C_\epsilon \in \mathcal{K}$, and $A_\epsilon \in \mathcal{A}_0$ such that

$$A_\epsilon \subset C_\epsilon \subset A \quad \text{and} \quad \mu(A - A_\epsilon) < \epsilon.$$

We call $\mathcal{K}$ is μ-approximates $\mathcal{A}_0$.

Lemma 2.1: Every compact finitely additive measure is countably additive.

Proof: Suppose $(\mathcal{X}, \mathcal{A}_0, \mu)$ is given. There exists a compact class $\mathcal{K}$ that is μ-approximates $\mathcal{A}_0$. Let $\{A_n\} \subset \mathcal{A}_0$ such that $A_n \searrow \emptyset$. We need to show that $\mu(A_n) \searrow 0$. For given $\epsilon > 0$, let $B_n \in \mathcal{A}_0$ and $C_n \in \mathcal{K}$, such that

$$B_n \subset C_n \subset A_n \quad \text{and} \quad \mu(A_n - B_n) < \frac{\epsilon}{2^n}.$$

Suppose $\mu(A_n)$ does not go to 0, i.e., for all n, $\mu(A_n) > \epsilon$. Since we know that

$$\mu\Big(A_n - \bigcap_{k=1}^{n} B_k\Big) = \mu\Big(\bigcap_{k=1}^{n} A_k\Big) - \mu\Big(\bigcap_{k=1}^{n} B_k\Big) < \frac{\epsilon}{2},$$

we conclude that for all n,

$$\mu\Big(\bigcap_{k=1}^{n} B_k\Big) > \frac{\epsilon}{2}.$$

Next, for all n

$$\bigcap_{k=1}^{n} B_k \neq \emptyset,$$

and hence, we have for all n

$$\bigcap_{k=1}^{n} C_k \neq \emptyset,$$

which implies

$$\bigcap_{k=1}^{\infty} C_k \neq \emptyset$$

since $C_k \in \mathcal{K}$ and $\mathcal{K}$ is compact. Therefore, it follows that

$$\bigcap_{k=1}^{\infty} A_k \supset \bigcap_{k=1}^{\infty} C_k \neq \emptyset$$

implies

$$\lim_{n\to\infty} A_n \neq \emptyset,$$

which is a contradiction.

2.2 Kolmogorov consistency theorem

Suppose $S \subset T$ is finite subset and Q_S is measure on $(\mathcal{X}_S, \mathcal{A}_S)$ satisfying consistency condition (2.1). Let $(\mathcal{X}_{\{t\}}, \mathcal{A}_{\{t\}}, Q_{\{t\}})$ be a compact probability measure space. For each $t \in T$, there exists a compact class $\mathcal{K}_t \subseteq \mathcal{A}_{\{t\}}$ and $\mathcal{K}_t$, Q_t approximates $\mathcal{A}_{\{t\}}$. Then, there exists a unique probability measure Q_0 on $(\mathcal{X}_T, \mathcal{A}_T)$ such that

$$\Pi_S : \mathcal{X}_T \to \mathcal{X}_S, \text{ and } Q_0 \circ \Pi_S^{-1} = Q_S.$$

Proof: Define

$$\mathcal{D} = \{C : C = \Pi_t^{-1}(K), K \in \mathcal{K}_t, t \in T\}.$$

Let

$$\{\Pi_{t_i}^{-1}(C_{t_i}), i = 1, 2 \ldots\}$$

be a countable family of sets and

$$B_t = \bigcup_{t_i = t} \Pi_{t_i}^{-1}(C_{t_i})$$

If the countable intersection of $\{B_t, t \in T\}$ is empty, then B_{t_0} is empty for some t_0. Since $\mathcal{K}_{t_0}$ is a compact class and all $C_{t_i} \in \mathcal{K}_{t_0}$, we get a finite set of t_i's ($t_i = t_0$). Let us call it J for which

$$\bigcup_{t_i \in J} C_{t_i} = \emptyset \Longrightarrow \bigcup_{t_i \in J} \Pi_{t_i}^{-1}(C_{t_i}) = \emptyset$$

Since $\mathcal{D}$ is a compact class, $\mathcal{K}$ as a countable intersections of sets in $\mathcal{D}$ is a compact finitely additive class. We shall show that Q_0 is a compact measure, i.e., $\mathcal{K}$ Q_0-approximates $\mathcal{C}_0$. Take $C \in \mathcal{C}_0$ and $\epsilon > 0$. For some $S \subset T$,

$$C = \Pi_S^{-1}(B).$$

Choose a rectangle

$$\prod_{t \in S}(A_t) \subset B$$

so that for $A_t \in \mathcal{A}_t$

$$Q_S(B - \prod_{t \in S} A_t) < \frac{\epsilon}{2}$$
$$Q_0(\Pi_S^{-1}(B) - \Pi_S^{-1}(\prod_{t \in S} A_t)) < \frac{\epsilon}{2}.$$

For each t, choose $K_t \in \mathcal{K}_t$ such that $K_t \subset A_t$ and

$$Q_t(A_t) < Q_t(K_t) + \frac{\epsilon}{\text{cardinality}(S)}.$$

Let

$$K = \Pi_{t \in S} K_t \text{ for } K_t \in \mathcal{K}_t.$$

Then, $K \subset C$ and

$$\begin{aligned} Q_0\Big(\Pi_S^{-1}(B) - \Pi_S^{-1}(\prod_{t \in S} K_t)\Big) &= Q_0\Big(\Pi_S^{-1}(B - \prod_{t \in S} A_t)\Big) \\ &\quad + Q_0\Big(\Pi_S^{-1}(\prod_{t \in S} A_t) - \Pi_S^{-1}(\prod_{t \in S} K_t)\Big) \\ &< \epsilon. \end{aligned}$$

Q_0 extends to a countable additive measure on $\sigma(\mathcal{C})$.

($\mathcal{X}_t = R$, R^d, or a complete separate metric space.)

Example: $T = N$, $\mathcal{X}_t = R$.

Suppose

$$Q_n = \bigotimes_{t \in \{1,2,\ldots,n\}} Q_{\{t\}}.$$

Then, there exists $\{X_n, n \in N\}$ of random variables defined on R^∞.

Example: $T = [0,1]$.

Let $\{C(t,s), t, s \in T\}$ be a set of real valued function with $C(t,s) = C(s,t)$ and

$$\sum_{t,s \in S} a_t a_s C(t,s) \geq 0$$

for S finite. ($\{a_t, t \in S\} \subset R$) Let Q_S be a probability measure with characteristic function for $\mathbf{t} \in R^d$

$$\phi_{Q_S}(\mathbf{t}) = \exp\left(-\frac{1}{2}\mathbf{t}' \sum_S \mathbf{t}\right), \tag{2.2}$$

where

$$\sum_S = \left(C(u,v)\right)_{u,v \in S}$$

Q_S satisfies the condition of Kolmogorov consistency theorem. Therefore, there exists $\{X_t, t \in T\}$, a family of random variables such that joint distribution of $\{X_t, t \in S\}$ is Q_S.

Example: Take $t, s \in [0,1]$ and $C(t,s) = \min(t,s)$.

$$\begin{aligned} C(t,s) &= \int_0^1 1_{[0,t]}(u) \cdot 1_{[0,s]}(u)du \\ &= \min(t,s) \end{aligned}$$

Let $S = \{t_1, \ldots, t_n\}$, $\{a_1, \ldots, a_n\} \subset R$. Then,

$$\begin{aligned} \sum_{i,j} a_i a_j C(t_i, t_j) &= \sum_{i,j} a_i a_j \int_0^1 1_{[0,t_i]}(u) \cdot 1_{[0,t_j]}(u)du \\ &= \int_0^1 \left(\sum_{i=1}^n a_i 1_{[0,t_i]}(u)\right)^2 du \\ &\geq 0 \end{aligned}$$

since $\sum\sum a_i a_j C(t_i, t_j)$ is non-negative definite. Therefore, there exists a Gaussian process with covariance $C(t,s) = \min(t,s)$.

2.3 The finite dimensional family for Brownian motion

Given covariance function $C(t,s) = C(s,t)$ with $s, t \in T$, and for all $\{a_1, \dots, a_k\}$ and $\{t_1, \dots, t_k\} \subset T$, such that

$$\sum_{k,j} a_k a_j C(t_k, t_j) \geq 0,$$

then there exists a Gaussian process $\{X_t, t \in T\}$ with $EX_t = 0$ for all t and $C(t,s) = E(X_t X_s)$.

Example: $C(t,s) = \min(t,s)$, and $T = [0,1]$.

$$\begin{aligned}\min(t,s) &= \int_0^1 1_{[0,t]}(u) \cdot 1_{[0,s]}(u)du \\ &= EX_t X_s\end{aligned}$$

There exists $\{X_t, t \in T\}$ such that $C(t,s) = E(X_t X_s) = \min(t,s)$.

Since X_t is Gaussian, we know that

$$X_t \in L^2(\Omega, \mathcal{F}, P).$$

Let, with $\overline{\mathrm{SP}}^{L^2}$ denoting closure in L_2 of the subspace generated by

$$M(X) = \overline{\mathrm{SP}}^{L^2}(X_t, t \in T),$$

Consider the map

$$I(X_t) = 1_{[0,t]}(u) \in L_2([0,1]) \text{ (Lebesgue measure).}$$

I is an isometry. Therefore, for $(t_1, \dots, t_n)$ with $t_1 \leq \dots \leq t_n$

$$\begin{aligned}I(X_{t_k} - X_{t_{k-1}}) &= I(X_{t_k}) - I(X_{t_{k-1}}) \text{ (because } I \text{ is a linear map)} \\ &= 1_{[0,t_k]}(u) - 1_{[0,t_{k-1}]}(u) \\ &= 1_{(t_{k-1},t_k]}(u)\end{aligned}$$

For $k \neq j$,

$$\begin{aligned}E(X_{t_k} - X_{t_{k-1}})(X_{t_j} - X_{t_{j-1}}) &= \int_0^1 1_{(t_{k-1},t_k]}(u) \cdot 1_{(t_{j-1},t_j]}(u)du \\ &= 0\end{aligned}$$

$X_{t_k} - X_{t_{k-1}}$ is independent of $X_{t_j} - X_{t_{j-1}}$ if $(t_{k-1}, t_k] \cap (t_{j-1}, t_j] = \emptyset$ because

$$(t_{k-1}, t_k] \cap (t_{j-1}, t_j] = \emptyset \Longrightarrow E(X_{t_k} - X_{t_{k-1}})(X_{t_j} - X_{t_{j-1}}) = 0.$$

$\{X_t, t \in T\}$ is an independent increment process.

Given $t_0 = 0, X_0 = 0, t_0 \leq t_1 \leq \ldots \leq t_n$, we have

$$P(X_{t_k} - X_{t_{k-1}} \leq x_k, k = 1, 2\ldots, n) = \prod_{k=1}^{n} \int_{-\infty}^{\infty} \frac{1}{\sqrt{2\pi(t_k - t_{k-1})}} e^{-\frac{y_k^2}{2(t_k - t_{k-1})}} dy_k.$$

Using transformation

$$\begin{aligned} Y_{t_1} &= X_{t_1} \\ Y_{t_2} &= X_{t_2} - X_{t_1} \\ \vdots &= \vdots \\ Y_{t_n} &= X_{t_n} - X_{t_{n-1}} \end{aligned}$$

we can compute the joint density of $(X_{t_1}, \ldots, X_{t_n})$.

$$f_{(X_{t_1},\ldots,X_{t_n})}(x_1, \ldots, x_n) = \frac{1}{\prod_{k=1}^{n} \sqrt{2\pi(t_k - t_{k-1})}} \exp\left[-\frac{1}{2}\sum_{k=1}^{n} \frac{(x_k - x_{k-1})^2}{(t_k - t_{k-1})}\right].$$

Define

$$p(t, x, B) = \frac{1}{\sqrt{2\pi t}} \int_B e^{-\frac{(y-x)^2}{t}} dy, \quad t \geq 0$$

and

$$\tilde{p}(t, s, x, B) = p(t - s, x, B).$$

Exercise: Prove for $0 = t_0 \leq t_1 \leq \ldots \leq t_n$

$$Q_{X_{t_1},\ldots,X_{t_n}}(B_1 \times \cdots \times B_n) = \int_{B_n} \cdots \int_{B_1} p(t_1 - t_0, 0, dy_1)$$
$$p(t_2 - t_1, y_1, dy_2) \cdots p(p(t_n - t_{n-1}, y_{n-1}, dy_n)).$$

Suppose we are given transition function $p(t, s, x, B)$ with $x \leq t$. Assume that $s \leq t \leq u$

$$p(u, s, x, B) = \int p(t, s, x, dy)p(s, t, y, B) \text{ (C-Kolmogorov condition).}$$

Then, for $0 = t_0 \leq t_1 \leq \ldots \leq t_n$

$$Q^x_{t_1,\ldots,t_n}(B_1 \times \cdots \times B_n) = \int_{B_n} \cdots \int_{B_1} p(t_1, t_0, x, dy_1)$$
$$p(t_2, t_1, y_1, dy_2) \cdots p(p(t_n, t_{n-1}, y_{n-1}, dy_n)).$$

(Use Fubini's theorem): For this consistent family, there exists a stochastic process with Q as finite dimensional distributions.

Exercise 2: Check that $\tilde{p}(t, s, x, B)$ satisfies the above condition.

$$Q^x(X_s \in B_1, X_t \in B_2) = \int_{B_1} Q^x(X_t \in B_2 | X_s = y) Q^x \circ X_s^{-1}(dy),$$

where

$$Q^x(X_{t_n} \in B_n | X_{t_1}, \ldots, X_{t_{n-1}}) = p(t_n, t_{n-1}, X_{t_{n-1}}, B_n) \text{ (Markov)}.$$

The Gaussian process with covariance

$$\min(t, s), \quad t, s, \in [0, 1]$$

has independent increments and is Markov.

Remarks: Consider
- $X(\omega) \in R^{[0,1]}$
- $C[0, 1] \notin \sigma(C(R^{[0,1]}))$

$C[0,1]$ is not measurable, but $C[0, 1]$ has $Q_0^*(C[0, 1]) = 1$ (outer measure).

2.4 Properties of Brownian motion

Definition 2.3: A Gaussian process is a stochastic process X_t, $t \in T$ for which any finite linear combination of $\{X_t\}$ has a Gaussian distribution.

Notation-wise, one can write $X \sim GP(m, K)$, meaning the random function X is distributed as a GP with mean function m and covariance function K.

Remark:

$$X \sim N(\mu_X, \sigma_X^2)$$
$$Y \sim N(\mu_Y, \sigma_Y^2)$$
$$Z = X + Y$$

Then, $Z \sim N(\mu_Z, \sigma_Z^2)$, where

$$\mu_Z = \mu_X + \mu_Y \text{ and } \sigma_Z^2 = \sigma_X^2 + \sigma_Y^2 + 2\rho_{X,Y}\sigma_X\sigma_Y.$$

Proposition 2.1: Given covariance $C(t,s) = C(s,t)$ with $s,t \in T$ and for all $\{a_1, \dots, a_k\} \subset R$ and $\{t_1, \dots, t_k\} \subset T$

$$\sum_{k,j} a_k a_j C(t_k, t_j) \geq 0,$$

then there exists a Gaussian process such that

$$\{X_t, t \in T\} \text{ with for all } t, EX_t = 0, C(t,s) = E(X_t, X_s)$$

Example: For $C(t,s) = \min(t,s)$ and $T = [0,1]$, we recall the properties of the associated Gaussian process.

In this example,

$$\begin{aligned} \min(t,s) &= \int_0^1 1_{[0,t]}(u) 1_{[0,s]}(u) du \\ &= EX_t X_s \end{aligned}$$

Thus, there exists $\{X_t, t \in T\}$, which is a Gaussian process, such that for all t

$$EX_t = 0, \text{ and } C(t,s) = E(X_t, X_s) = \min(t,s).$$

Since X_t is Gaussian, we know that

$$X_t \in L^2(\Omega, \mathcal{F}, P).$$

Let

$$M(X) = \overline{SP}^{L^2}(X_t, t \in T) \text{ (SP means “span”).}$$

Consider the map $I : M(X) \to \overline{SP}\{1_{[0,t](u)}, t \in [0,1]\}$ such that

$$I(X_t) = 1_{[0,t]}(u)$$

and

$$I\left(\sum a_k X_{t_k}\right) = \sum a_k I(X_{t_k}).$$

Proposition 2.2: I is a linear map.

Exercise: Prove Proposition 2.2.

Proposition 2.3: I is an isometry

Proof: Since $\{X_t, t \in T\}$ is Gaussian,

$$\begin{aligned} Var(X_{t_k} - X_{t_{k-1}}) &= Var(X_{t_k}) + Var(X_{t_k}) - 2Cov(X_{t_k}, X_{t_{k-1}}) \\ &= C(t_k, t_k) + C(t_{k-1}, t_{k-1}) - 2C(t_{k-1}, t_k) \\ &= t_k + t_{k-1} - 2t_{k-1} \\ &= t_k - t_{k-1}. \end{aligned}$$

Therefore,

$$\begin{aligned} ||X_{t_k} - X_{t_{k-1}}||_{L_2}^2 &= \int_{[0,1]} (X_{t_k} - X_{t_{k-1}})^2 dP \\ &= E(X_{t_k} - X_{t_{k-1}})^2 \\ &= Var(X_{t_k} - X_{t_{k-1}}) \\ &= t_k - t_{k-1}. \end{aligned}$$

Also,

$$\begin{aligned} ||I(X_{t_k}) - I(X_{t_{k-1}})||_{L_2}^2 &= ||1_{[0,t_k]}(u) - 1_{[0,t_{k-1}]}(u)||_{L_2}^2 \\ &= ||1_{(t_{k-1},t_k]}(u)||_{L_2}^2 \\ &= t_k - t_{k-1} \\ &= ||X_{t_k} - X_{t_{k-1}}||_{L_2}^2. \end{aligned}$$

This completes the proof.

Exercise: I can be extended by continuity on $M(X)$ onto $L^2([0,1], \lambda)$

We call $I^{-1}(f)$ the stochastic integral of f with respect to Brownian motion for $f \epsilon L^2([0,1], \lambda)$.

Suppose that $t_2 \le \ldots \le t_k$. Then,

$$\begin{aligned} I(X_{t_k} - X_{t_{k-1}}) &= I(X_{t_k}) - I(X_{t_{k-1}}) \text{ (since } I \text{ is an linear map)} \\ &= 1_{[0,t_k]}(u) - 1_{[0,t_{k-1}]}(u) \\ &= 1_{(t_{k-1},t_k]}(u). \end{aligned}$$

$X_{t_k} - X_{t_{k-1}}$ is independent of $X_{t_j} - X_{t_{j-1}}$ if

$$(t_{k-1}, t_k] \cap (t_{j-1}, t_j] = \emptyset.$$

Proposition 2.4: If $X_{t_k} - X_{t_{k-1}}$ is independent of $X_{t_j} - X_{t_{j-1}}$, then

$$E(X_{t_k} - X_{t_{k-1}})(X_{t_j} - X_{t_{j-1}}) = 0.$$

Proof: For $k \neq j$,

$$\begin{aligned} & E(X_{t_k} - X_{t_{k-1}})(X_{t_j} - X_{t_{j-1}}) \\ &= \int_0^1 1_{(t_{k-1},t_k]}(u) 1_{(t_{j-1},t_j]}(u) du \\ &= 0. \end{aligned}$$

Suppose $\{X_t, t \in T\}$ is an independent increment process such that $t_0 = 0, X_0 = 0$, and $t_0 \leq t_1 \leq \ldots \leq t_n$. Then, $X_{t_k} - X_{t_{k-1}}$ is Gaussian with mean 0 and variance $t_k - t_{k-1}$.

$$\begin{aligned} P(X_{t_k} - X_{t_{k-1}} \leq x_k, k = 1, 2\ldots, n) &= \prod_{k=1}^{n} P(X_{t_k} - X_{t_{k-1}} \leq x_k) \\ &= \prod_{k=1}^{n} \int_{-\infty}^{x_k} \frac{1}{\sqrt{2\pi(t_k - t_{k-1})}} e^{-\frac{y_k^2}{2(t_k - t_{k-1})}} dy_k. \end{aligned}$$

Let

$$\begin{aligned} Y_{t_1} &= X_{t_1} \\ Y_{t_2} &= X_{t_2} - X_{t_1} \\ \vdots &\quad \vdots \\ Y_{t_n} &= X_{t_n} - X_{t_{n-1}}. \end{aligned}$$

Then,

$$f_{X_{t_1},\ldots,X_{t_n}}(x_1, \ldots, x_n) = \frac{1}{\prod_{k=1}^{n} \sqrt{2\pi(t_k - t_{k-1})}} \exp\left[-\frac{1}{2}\sum_{k=1}^{n} \frac{(x_k - x_{k-1})^2}{(t_k - t_{k-1})}\right].$$

2.5 Kolmogorov continuity theorem

For each t, if $P(\tilde{X}_t = X_t) = 1$, then we say the finite dimensional distributions of $\tilde{X}_t$ and X_t are the same. We call $\tilde{X}_t$ a version of X_t.

Proposition 2.5: Let $\{X_t, t \in [0, 1]\}$ be a stochastic process with

$$E|X_t - X_s|^{\beta} \leq C|t - s|^{1+\alpha} \text{ with } C, \alpha, \beta > 0.$$

Then, there exists a version of $\{X_t, t \in [0, 1]\}$, which has a continuous sample paths.

Corollary 2.1: The Gaussian process with covariance function

$$C(t, s) = \min(t, s), \ t, s \in [0, 1]$$

(has independent increment and is Markovian) has a version that is continuous.

$$\begin{aligned} E(X_t - X_s)^2 &= EX_t^2 - 2EX_tX_s + EX_s^2 \\ &= t - 2s + s \\ &= |t - s| \\ E(X_t - X_s)^4 &= 3[E(X_t - X_s)^2]^2 \\ &= 3|t - s|^2 \end{aligned}$$

We shall denote the continuous version by W_t.

Proof of Proposition 2.5: Take $0 < \gamma < \frac{\alpha}{\beta}$ and $\delta > 0$ such that

$$(1 - \delta)(1 + \alpha - \beta\gamma) > 1 + \delta.$$

For $0 \le i \le j \le 2^n$ and $|j - i| \le 2^{n\delta}$,

$$\begin{aligned} \sum_{i,j} P\left(|X_{j2^{-n}} - X_{i2^{-n}}| > [(j - i)2^{-n}]^\gamma\right) &\le C \sum_{i,j} [(j - i)2^{-n}]^{-\beta\gamma + (1+\delta)} \text{ (by Chevyshev)} \\ &= C_2 2^{n[(1+\delta)-(1-\delta)(1+\alpha-\beta\gamma)]} \\ &< \infty, \end{aligned}$$

where $(1 + \delta) - (1 - \delta)(1 + \alpha - \beta\gamma) = -\mu$.

Then, by the Borell-Cantelli lemma,

$$P\left(|X_{j2^{-n}} - X_{i2^{-n}}| > [(j - i)2^{-n}]^\gamma\right) = 0,$$

i.e., there exists $n_0(\omega)$ such that for all $n \ge n_0(\omega)$

$$|X_{j2^{-n}} - X_{i2^{-n}}| \le [(j - i)2^{-n}]^\gamma.$$

Let $t_1 < t_2$ be rational numbers in $[0, 1]$ such that

$$\begin{aligned} &t_2 - t_1 \le 2^{-n_0(1-\delta)} \\ &t_1 = i2^{-n} - 2^{-p_1} - \ldots - 2^{-p_k} \ (n < p_1 < \ldots < p_k) \\ &t_2 = j2^{-n} - 2^{-q_1} - \ldots - 2^{-q_k} \ (n < q_1 < \ldots < q_k) \\ &t_1 \le i2^{-n} \le j2^{-n} \le t_2. \end{aligned}$$

Let

$$h(t) = t^\gamma \text{ for } 2^{-(n+1)(1-\delta)} \le t \le 2^{-n(1-\delta)}.$$

Then,

$$\left|X_{t_1} - X_{i2^{-n}}\right| \le C_1 h(2^{-n})$$

$$\left|X_{t_2} - X_{j2^{-n}}\right| \le C_2 h(2^{-n})$$

$$\left|X_{t_2} - X_{t_1}\right| \le C_3 h(t_2 - t_1)$$

and

$$\left|X_{i2^{-n}-2^{-p_1}-\ldots-2^{-p_k}} - X_{i2^{-n}-2^{-p_1}-\ldots-2^{-p_{k-1}}}\right| \le 2^{-p_k \gamma}.$$

Under this condition, the process $\{X_t, t \in [0,1]\}$ is uniformly continuous on rational numbers in $[0,1]$.

Let

$$\psi : (\Omega_Q, \mathcal{C}_{\Omega_Q}) \to (C[0,1], \sigma(C[0,1])),$$

which extends uniformly continuous functions in rational to continuous function in $[0,1]$.

Let P is a measure generated by the same finite dimensional on rationals. Then

$$\tilde{P} = P \circ \psi^{-1}$$

is the measure of $\tilde{X}_t$, version of X_t.

For $\{X_t, t \in [0,1]\}$, there exists a version of continuous sample path. In case of Brownian motion, there exists a continuous version. We call it $\{W_t\}$.

$\{W_{t+t_0} - W_{t_0}, t \in [0,\infty]\}$ is a Weiner process.

2.6 Exit time for Brownian motion and Skorokhod theorem

It is well known that if $X_1, \ldots, X_n$ are independent identically distributed (iid) random variables with $EX_1 = 0$ and $EX_1^2 < \infty$, then

$$Y_n = \frac{X_1 + \ldots + X_n}{\sqrt{n}}$$

converges in distribution to standard normal random variable Z. This is equivalent to weak convergence of Y_n to Z, i.e., $Ef(Y_n) \to Ef(Z)$ for all bounded continuous functions for $\mathbb{R}$. If we denote by

$$Y_{n,m} = \frac{X_1 + \ldots + X_m}{\sqrt{n}} \text{ and } Y_{n,0} = 0$$

and define interpolated continuous process

$$Y_{n,t} = \begin{cases} Y_{n,m} \; t = m \epsilon \{0, 1, \ldots n\} \\ \text{linear if } t \epsilon [m-1, m] \end{cases}$$

for $m = 0, 1, 2, \ldots n$ and we shall show that $Ef(Y_{n,[n.]} \to Ef(W.)$ for all bounded continuous functions for $C[0,1]$ with $\{W(t), t\epsilon[0, 1]\}$ Brownian motion. In fact, we shall use Skorokhod embedding to prove for any $\epsilon > 0$

$$P(\sup_{0 \le t \le 1} |Y_{n,[nt]} - W(t)| > \epsilon) \to 0$$

as $n \to \infty$, which implies the above weak convergences (see Theorem 2.3).

Let

$$\mathcal{F}_t^W = \sigma(W_s, s \le t).$$

τ is called "stopping time" if

$$\{\tau \le t\} \in \mathcal{F}_t^W.$$

Define

$$\mathcal{F}_\tau = \{A : A \cap \{\tau \le t\} \in \mathcal{F}_t^W\}.$$

Then $\mathcal{F}_\tau$ is a σ-field. $\frac{W_t}{\sqrt{t}}$ is a standard normal variable.
Define

$$T_a = \inf\{t : W_t = a\}.$$

Then, $T_a < \infty$ a.e. and is a stopping time.

Theorem 2.1: Let $a < x < b$. Then

$$P_x(T_a < T_b) = \frac{b-x}{b-a}.$$

Remark: W_t is Gaussian and has independent increment. Also, for $s \le t$

$$E(W_t - W_s | \mathcal{F}_s) = 0$$

and hence $\{W_t\}$ is martingale.

Proof: Let $T = T_a \wedge T_b$. We know $T_a, T_b < \infty$ a.e., and

$$W_{T_a} = a, \text{ and } W_{T_b} = b.$$

Since $\{W_t\}$ is MG,

$$\begin{aligned} E_x W_T &= E_x W_0 \\ &= x \\ &= aP(T_a < T_b) + b(1 - P(T_a < T_b)). \end{aligned}$$

Therefore,

$$P_x(T_a < T_b) = \frac{b - x}{b - a}.$$

$\{W_t, t \in [0, \infty]\}$ is a Weiner Process, starting from x with $a < x < b$. We know that

$$\begin{aligned} E((W_t - W_s)^2 | \mathcal{F}_s^W) &= E(W_t - W_s)^2 \\ &= (t - s). \end{aligned}$$

Also,

$$\begin{aligned} E((W_t - W_s)^2 | \mathcal{F}_s^W) &= E(W_t^2 | \mathcal{F}_s^W) - 2E(W_t W_s | \mathcal{F}_s^W) + E(W_s^2 | \mathcal{F}_s^W) \\ &= E(W_t^2 | \mathcal{F}_s^W) - W_s^2 \\ &= E(W_t^2 - W_s^2 | \mathcal{F}_s^W) \\ &= (t - s). \end{aligned}$$

Therefore,

$$E(W_t^2 - t | \mathcal{F}_s^W) = W_s^2 - s,$$

and hence $\{(W_t^2 - t), t \in [0, \infty]\}$ is a martingale.

Suppose that $x = 0$ and $a < 0 < b$. Then $T = T_a \wedge T_b$ is a finite stopping time. Therefore,

$$T \wedge t$$

is also stopping time.

$$E(W_{T \wedge t}^2 - T \wedge t) = 0$$

$$E_0(W_T^2) = E_0 T$$

$$\begin{aligned} EW_T^2 &= ET \\ &= a^2 P(T_a < T_b) + b^2(1 - P(T_a < T_b)) \\ &= -ab \end{aligned}$$

Suppose X has two values a, b with $a < 0 < b$ and

$$\begin{aligned} EX &= aP(X = a) + bP(X = b) \\ &= 0 \end{aligned}$$

Remark:

$$P(X = a) = \frac{b}{b - a} \text{ and } P(X = b) = -\frac{a}{b - a}$$

Let $T = T_a \wedge T_b$. Then, W_T has the same distribution as X. We denote

$$\mathcal{L}(W_T) = \mathcal{L}(X)$$

or

$$W_T =_{\mathcal{D}} X$$

2.6.1 Skorokhod theorem

Let X be random variable with $EX = 0$ and $EX^2 < \infty$. Then, there exists a $\mathcal{F}_t^W$-stopping time T such that

$$\mathcal{L}(W_T) = \mathcal{L}(X) \text{ and } ET = EX^2.$$

Proof: Let $F(x) = P(X \le x)$.

$$\begin{aligned} EX = 0 &\Rightarrow \int_{-\infty}^{0} u dF(u) + \int_{0}^{\infty} v dF(v) = 0 \\ &\Rightarrow -\int_{-\infty}^{0} u dF(u) = \int_{0}^{\infty} v dF(v) = C. \end{aligned}$$

Let ψ be a bounded function with $\psi(0) = 0$. Then,

$$\begin{aligned} &C \int_R \psi(x) dF(x) \\ &= C\Big(\int_0^\infty \psi(v) dF(v) + \int_{-\infty}^0 \psi(u) dF(u) \Big) \\ &= \int_0^\infty \psi(v) dF(v) \int_{-\infty}^0 -u dF(u) + \int_{-\infty}^0 \psi(u) dF(u) \int_0^\infty v dF(v) \\ &= \int_0^\infty dF(v) \int_{-\infty}^0 dF(u)(v\psi(u) - u\psi(v)). \end{aligned}$$

Therefore,

$$\begin{aligned}\int_R \psi(x)dF(x) &= C^{-1}\int_0^\infty dF(v)\int_{-\infty}^0 dF(u)(v\psi(u) - u\psi(v))\\ &= C^{-1}\int_0^\infty dF(v)\int_{-\infty}^0 dF(u)(v-u)\left[\frac{v}{v-u}\psi(u) - \frac{u}{v-u}\psi(v)\right].\end{aligned}$$

Consider (U, V) be a random vector in R^2 such that

$$P\big[(U, V) = (0, 0)\big] = F(\{0\})$$

and for $A \subset (-\infty, 0) \times (0, \infty)$

$$P((U, V) \in A) = C^{-1}\int\int_A dF(u)dF(v)(v-u).$$

If $\psi = 1$,

$$\begin{aligned}&P((U, V) \in (-\infty, 0) \times (0, \infty))\\ &= C^{-1}\int_0^\infty dF(v)\int_{-\infty}^0 dF(u)(v-u)\\ &= C^{-1}\int_0^\infty dF(v)\int_{-\infty}^0 dF(u)(v-u)\left[\frac{v}{v-u}\psi(u) - \frac{u}{v-u}\psi(v)\right]\\ &= \int_R \psi(x)dF(x)\\ &= \int_R dF(x)\\ &= 1,\end{aligned}$$

and hence, P is a probability measure.

Let $u < 0 < v$ such that

$$\mu_{U,V}(\{u\}) = \frac{v}{v-u} \text{ and } \mu_{U,V}(\{v\}) = -\frac{u}{v-u}.$$

Then, by Fubini,

$$\int \psi(x)dF(x) = E\int \psi(x)\mu_{u,v}(dx).$$

On product space $\Omega \times \Omega'$, let

$$\begin{aligned}W_t(\omega, \omega') &= W_t(\omega)\\ (U, V)(\omega, \omega') &= (U, V)(\omega').\end{aligned}$$

$T_{U,V}$ is not a stopping time on $\mathcal{F}_t^W$.

We know that if $U = u$ and $V = v$

$$\mathcal{L}(T_{U,V}) = \mathcal{L}(X).$$

Then,

$$\begin{aligned} ET_{U,V} &= E_{U,V}E(T_{U,V}|U, V) \\ &= -EUV \\ &= C^{-1}\int_{-\infty}^{0} dF(u)(-u)\int_{0}^{\infty} dF(v)v(v-u) \\ &= -\int_{0}^{\infty} dF(v)(-u)\left[C^{-1}\int_{0}^{\infty} vdF(v) - u\right] \\ &= EX^2. \end{aligned}$$

2.7 Embedding of sums of i.i.d. random variable in Brownian motion

Let $t_0 \in [0, \infty)$. Then,

$$\{W(t+t_0) - W(t_0), t \geq 0\}$$

is a Brownian motion and independent of $\mathcal{F}_{t_0}$.
Let τ be a stopping time w.r.t. $\mathcal{F}_t^W$. Then, one can see that

$$W_t^*(\omega) = W_{\tau(\omega)+t}(\omega) - W_{\tau(\omega)}(\omega)$$

is a Brownian motion w.r.t. $\mathcal{F}_\tau^W$ where

$$\mathcal{F}_\tau^W = \{B \in \mathcal{F} : B \cap \{\tau \leq t\} \in \mathcal{F}_t^W\} \text{ for all } t \leq 0\}.$$

Let V_0 be countable. Then,

$$\{\omega : W_*^t(\omega) \in B\} = \bigcup_{t_0 \in V_0} \{\omega : W(t+t_0) - W(t_0) \in B, \tau = t_0\}.$$

For $A \in \mathcal{F}_\tau^W$,

$$\begin{aligned} P[\{(W_{t_1}^*, \ldots, W_{t_k}^*) \in B_k\} \cap A] &= \sum_{t_0 \in V_0} P[\{(W_{t_1}, \ldots, W_{t_k}) \in B_k\} \cap A \cap \{\tau = t_0\}] \\ &= \sum_{t_0 \in V_0} P\Big((W_{t_1}^*, \ldots, W_{t_k}^*) \in B_k\Big) \cdot P(A \cap \{\tau = t_0\}) \end{aligned}$$

(because of independence)

$$= P\Big((W^*_{t_1}, \ldots, W^*_{t_k}) \in B_k\Big) \sum_{t_0 \in V_0} P(A \cap \{\tau = t_0\})$$
$$= P\Big((W^*_{t_1}, \ldots, W^*_{t_k}) \in B_k\Big) P(A)$$

Let τ be any stopping time such that

$$\tau_n = \begin{cases} 0, & \text{if } \tau = 0; \\ \frac{k}{2^n}, & \text{if } \frac{k-1}{2^n} \le \tau < \frac{k}{2^n}. \end{cases}$$

If $\frac{k}{2^n} \le t < \frac{k+1}{2^n}$,

$$\{\tau_n \le t\} = \left\{\tau \le \frac{k}{2^n}\right\} \in \mathcal{F}_{\frac{k}{2^n}} \subset \mathcal{F}_t.$$

Claim: $\mathcal{F}_\tau \subset \mathcal{F}_{\tau_n}$.

Proof: Suppose $C \in \mathcal{F}_\tau = \{B : B \cap \{\tau \le t\} \in \mathcal{F}_t\}$. Then, since $\frac{k}{2^n} \le t$,

$$C \cap \{\tau \le t\} = C \cap \left\{\tau \le \frac{k}{2^n}\right\} \in \mathcal{F}_{\frac{k}{2^n}} \in \mathcal{F}_t.$$

This completes the proof.

$W^n_{t_1} = W_{\tau_n + t} - W_{\tau_n}$ is a Brownian Motion for each n, independent of $\mathcal{F}_\tau$.

Theorem 2.2: Let $X_1, \ldots, X_n$ be i.i.d. with $EX_i = 0$, $EX_i^2 < \infty$ for all i. Then, there exists a sequence of stopping time $T_0 = 0, T_1, \ldots, T_n$ such that

$$\mathcal{L}(S_n) = \mathcal{L}(W_{T_n}),$$

where $(T_i - T_{i-1})$ are i.i.d.

Proof: $(U_1, V_1), \ldots, (U_n, V_n)$ i.i.d. as (U, V) and independent of W_t.

$$T_0 = 0, T_k = \inf\{t \ge T_{k-1}, W_{t+T_{k-1}} - W_{T_{k-1}} \in (U_k, V_k)\}.$$

$T_k - T_{k-1}$ are i.i.d.

$$\mathcal{L}(X_1) = \mathcal{L}(W_{T_1})$$
$$\mathcal{L}(X_2) = \mathcal{L}(W_{T_2} - W_{T_1})$$
$$\vdots \quad \vdots \quad \vdots$$
$$\mathcal{L}(X_n) = \mathcal{L}(W_{T_n} - W_{T_{n-1}}).$$

Then,

$$\begin{aligned}
\mathcal{L}\Big(\frac{S_n}{\sqrt{n}}\Big) &= \mathcal{L}\Big(\frac{W(T_n)}{\sqrt{n}}\Big) \\
&= \mathcal{L}\Big(\frac{W(T_n/n \cdot n)}{\sqrt{n}}\Big) \\
&= \mathcal{L}\Big(W\Big(\frac{T_n}{n}\Big)\Big) \quad \Big(\text{since } \frac{W(nt)}{\sqrt{n}} \approx W(t)\Big).
\end{aligned}$$

Assume $EX_1^2 = 1$. Then

$$\frac{T_n}{n} \to_{a.s.} E(T_1) = EX_1^2 = 1,$$

and hence

$$\frac{S_n}{\sqrt{n}} \to W_1.$$

2.8 Donsker's theorem

$X_{n,1}, \ldots, X_{n,n}$ are i.i.d. for each n with $EX_{n,m} = 0$, $EX_{n,m}^2 < \infty$, and $S_{n,m} = X_{n,1} + \cdots + X_{n,m} = W_{\tau_m^n}$, where τ_m^n is stopping time and W is Brownian motion. Define

$$S_{n,u} = \begin{cases} S_{n,m}, & \text{if } u = m \in \{0, 1, 2, \ldots, n\}; \\ \text{linear}, & \text{if } u \in [m-1, m]. \end{cases}$$

Lemma 2.2: If $\tau_{[ns]}^n \to s$ for $s \in [0, 1]$, then with sup norm $|| \; ||_\infty$ on $C[0, 1]$,

$$||S_{n,[n\cdot]} - W(\cdot)||_\infty \to 0 \text{ in probability.}$$

Proof: For given $\epsilon > 0$, there exists $\delta > 0 (1/\delta$ is an integer) such that

$$P\Big(|W_t - W_s| < \epsilon, \text{ for all } t, s \in [0, 1], |t - s| < 2\delta\Big) > 1 - \epsilon \tag{2.3}$$

τ_m^n is increasing in m. For $n \geq N\delta$,

$$P\Big(|\tau_{nk\delta}^n - k\delta| < \delta, k = 1, 2, \ldots, \frac{1}{\delta}\Big) \geq 1 - \epsilon$$

since $\tau_{[ns]}^n \to s$. For $s \in ((k-1)\delta, k\delta)$, we have

$$\tau_{[ns]}^n - s \geq \tau_{[n(k-1)\delta]}^n - k\delta$$

$$\tau^n_{[ns]} - s \le \tau^n_{[nk\delta]} - (k-1)\delta.$$

Combining these, we have for $n \ge N\delta$

$$P\Big(\sup_{0\le s\le 1} |\tau^n_{[ns]} - s| < 2\delta\Big) > 1 - \epsilon. \tag{2.4}$$

For ω in event in (2.3) and (2.4), we get for $m \le n$, as $W_{t^n_m} = S_{n,m}$,

$$\Big|W_{\tau^n_m} - W_{\frac{m}{n}}\Big| < \epsilon.$$

For $t = \frac{m+\theta}{n}$ with $0 < \theta < 1$,

$$\Big|S_{n,[nt]} - W_t\Big| \le (1-\theta)\Big|S_{n,m} - W_{\frac{m}{n}}\Big| + \theta\Big|S_{n,m+1} - \frac{W_{m+1}}{n}\Big|$$
$$+(1-\theta)\Big|W_{\frac{m}{n}} - W_t\Big| + \theta\Big|\frac{W_{n+1}}{n} - W_t\Big|.$$

For $n \ge N_\delta$ with $\frac{1}{n} < 2\delta$,

$$P\Big(||S_{n,[ns]} - W_s||_\infty \ge 2\epsilon\Big) < 2\epsilon.$$

Theorem 2.3: Let f be bounded and continuous function on [0,1]. Then

$$Ef\Big(S_{n,[n\cdot]}\Big) \to Ef(W(\cdot)).$$

Proof: For fixed $\epsilon > 0$, define

$$G_\delta = \{W, W' \in C[0,1] : ||W - W'||_\infty < \delta \text{ implies } |f(W) - f(W')| < \epsilon\}.$$

Observe that $G_\delta \uparrow C[0,1]$ as $\delta \downarrow 0$. Then,

$$\Big|Ef\Big(S_{n,[n\cdot]}\Big) - Ef(W(\cdot))\Big| \le \epsilon + 2M\Big(P(G^c_\delta) + P\Big(||S_{n,[n\cdot]} - W(\cdot)|| > \delta\Big)\Big).$$

Since $P(G^c_\delta) \to 0$ and $P\Big(||S_{n,[n\cdot]} - W(\cdot)|| > \delta\Big) \to 0$ by Lemma 2.2.

For

$$f(x) = \max_t |x(t)|,$$

we have

$$\max_t \left|\frac{S_{[nt]}}{\sqrt{n}}\right| \to \max_t |W(t)| \text{ in distribution}$$

and

$$\max_{1\le m\le n}\left|\frac{S_m}{\sqrt{n}}\right| \to \max_t |W(t)| \text{ in distribution.}$$

Let

$$R_n = 1 + \max_{1\le m\le n} S_m - \min_{1\le m\le n} S_m$$

Then

$$\frac{R_n}{\sqrt{n}} \Rightarrow_{weakly} \max_{0\le t\le 1} W(t) - \min_{0\le t\le 1} W(t).$$

We now derive from Donsker theorem invariance principle for U-statistics

$$\prod_{i=1}^{[nt]}\left(1+\frac{\theta X_i}{\sqrt{n}}\right) = \sum_{k=1}^{[nt]} n^{-\frac{k}{2}} \sum_{1\le i_1\le\cdots\le i_k\le n} X_{i_1}\cdots X_{i_k},$$

where X_i are i.i.d. and $EX_i^2 < \infty$. Next, using CLT, SLLN, and the fact $P(max|\frac{X_i}{\sqrt{n}}| > \epsilon) \to 0$,

$$\begin{aligned}
\log\left[\prod_{i=1}^{[nt]}\left(1+\frac{\theta X_i}{\sqrt{n}}\right)\right] &= \sum_{i=1}^{[nt]} \log\left(1+\frac{\theta X_i}{\sqrt{n}}\right) \\
&= \theta\sum_{i=1}^{[nt]}\frac{X_i}{\sqrt{n}} - \frac{\theta^2}{2}\sum_{i=1}^{[nt]}\frac{X_i^2}{n} + \frac{\theta^3}{3}\sum_{i=1}^{[nt]}\frac{X_i^3}{n\sqrt{n}} - \cdots \\
&\Rightarrow \theta W(t) - \frac{\theta^2}{2}t,
\end{aligned}$$

and hence

$$\prod_{i=1}^{[nt]}\left(1+\frac{\theta X_i}{\sqrt{n}}\right) \Rightarrow e^{\theta W(t) - \frac{\theta^2}{2}t}.$$

2.9 Empirical distribution function

Let us define empirical distribution function

$$\hat{F}_n(x) = \frac{1}{n}\sum_{i=1}^{n} \mathbf{1}(X_i \le x), \quad x \in \mathbb{R}.$$

Then Glivenko-Cantelli lemma says

$$\sup_x |\hat{F}_n(x) - F(x)| \to_{a.s.} 0.$$

Assume F is continuous. Let $U_i = F(X_i)$. For $y \in [0, 1]$, define

$$\hat{G}(y) = \frac{1}{n}\sum_{i=1}^{n} \mathbf{1}(F(X_i) \leq y).$$

Then by 1-1 transformation,

$$\sqrt{n}\sup_x |\hat{F}_n(x) - F(x)| = \sqrt{n}\sup_{y\in[0,1]} |\hat{G}_n(y) - y|.$$

Let $U_1, U_2, \dots, U_n$ be uniform distribution and let $U_{(i)}$ be order statistic such that

$$U_{(1)}(\omega) \leq \cdots \leq U_{(n)}(\omega).$$

Next,

$$f\Big(U_{(1)}, \dots, U_{(n)}\Big) = f\Big(U_{\pi(1)}, \dots, U_{\pi(n)}\Big)$$

$$\begin{aligned} f_{U_{\pi(1)},\dots,U_{\pi(n)}}(u_1, \dots, u_n) &= f_{U_1,\dots,U_n}(u_1, \dots, u_n) \text{ if } u_1 < u_2 < \dots < u_n. \\ &= \begin{cases} 1, & \text{if } \mathbf{u} \in [0, 1]^n; \\ 0, & \text{if } \mathbf{u} \notin [0, 1]^n. \end{cases} \end{aligned}$$

For bounded g,

$$\begin{aligned} Eg\Big(U_1, \dots, U_n\Big) = \sum_{\pi\in\Pi} \int_{u_1<u_2<\dots<u_n} & g^{(u_1\dots u_n)}_{U_{\pi(1)},\dots,\pi(2),\dots,U_{\pi(n)}}(u_1, \dots, u_n) \\ & \times f\Big(u_{\pi(1)}, \dots, u_{\pi(n)}\Big) du_1 \cdots du_n. \end{aligned}$$

So we get

$$f_{U_1,\dots,U_n}(u_1, \dots, u_n) = \begin{cases} n!, & \text{if } u_1 < u_2 < \cdots < u_n; \\ 0, & \text{otherwise.} \end{cases}$$

Theorem 2.4: Let e_j be i.i.d. exponential distribution with failure rate λ. Then

$$\mathcal{L}\Big(U_1, \dots, U_n\Big) = \mathcal{L}\bigg(\frac{Z_1}{Z_{n+1}}, \dots, \frac{Z_n}{Z_{n+1}}\bigg),$$

where

$$Z_i = \sum_{j=1}^{i} e_j.$$

Proof: First we have

$$f_{e_1,\ldots,e_{n+1}}(u_1,\ldots,u_{n+1}) = \begin{cases} \lambda^{n+1}e^{-\sum_{i=1}^{n+1} u_i\lambda}, & \text{if } u_i \geq 0; \\ 0, & \text{otherwise.} \end{cases}$$

Let $s_i - s_{i-1} = u_i$ for $i = 1, 2, \ldots, n+1$. Then

$$f_{Z_1,\ldots,Z_{n+1}}(s_1,\ldots,s_{n+1}) = \prod_{i=1}^{n+1} \lambda e^{-(s_i - s_{i-1})\lambda}.$$

Let

$$\begin{aligned} v_i &= \frac{s_i}{s_{n+1}} \text{ for } i \leq n \\ v_{n+1} &= s_{n+1}. \end{aligned}$$

Then

$$\begin{aligned} f_{V_1,\ldots,V_{n+1}}(v_1,\ldots,v_{n+1}) &= \prod_{i=1}^{n} \left(\lambda e^{-\lambda v_{n+1}(v_i - v_{i-1})}\right)\lambda e^{-\lambda v_{n+1}(1-v_n)} \\ &= \lambda^{n+1} e^{-\lambda v_n v_{n+1} - \lambda v_{n+1} + \lambda v_n v_{n+1}} v_{n+1}^n \end{aligned}$$

Exercise: Integrate with respect to v_{n+1} to complete the proof.

$$\begin{aligned} D_n &= \sqrt{n} \max_{1\leq m\leq n} \left| \frac{Z_m}{Z_{n+1}} - \frac{m}{n} \right| \\ &= \frac{n}{Z_{n+1}} \max_{1\leq m\leq n} \left| \frac{Z_m}{\sqrt{n}} - \frac{m}{n}\frac{Z_{n+1}}{\sqrt{n}} \right| \\ &= \frac{n}{Z_{n+1}} \max_{1\leq m\leq n} \left| \frac{Z_m - m}{\sqrt{n}} - \frac{m}{n}\frac{Z_{n+1} - n}{\sqrt{n}} \right| \\ &= \frac{n}{Z_{n+1}} \max_{1\leq m\leq n} \left| W_n(t) - t\left(W_n(1) + \frac{Z_{n+1} - Z_n}{\sqrt{n}}\right) \right|. \end{aligned}$$

where

$$W_n(t) = \begin{cases} \frac{Z_m - m}{\sqrt{n}}, & \text{if } t = \frac{m}{n}; \\ \text{linear}, & \text{between.} \end{cases}$$

We know that $n/Z_{n+1} \to_{a.s.} \lambda$ and

$$E\left(\frac{Z_{n+1} - Z_n}{\sqrt{n}}\right)^2 = \frac{1}{n} E e_n^2 \to 0,$$

and hence by Chevyshev's inequality,

$$\frac{Z_{n+1} - Z_n}{\sqrt{n}} \to_p 0.$$

Since max(·) is a continuous function and

$$\left(W_n(\cdot) - \cdot W_n(1)\right) \Rightarrow_D \left(W(\cdot) - \cdot W(1)\right),$$

we have for $\lambda = 1$

$$\begin{aligned} D_n &= \underbrace{\frac{n}{Z_{n+1}}}_{\to_{a.s}1} \max_{1\le m\le n} \left| W_n(t) - t\Big(W_n(1) + \underbrace{\frac{Z_{n+1} - Z_n}{\sqrt{n}}}_{\to_p 0}\Big)\right| \\ &\Rightarrow_D \max_{0\le t\le 1} \left|W(t) - tW(1)\right|. \end{aligned}$$

The process $\{W(t) - tW(1), 0 \le t \le 1\}$ is called Brownian Bridge.

Observe,

$$\begin{aligned} &P\left(W_{t_1} \le x_1, W_{t_2} \le x_2, \ldots, W_{t_k} \le x_k, W_1 = 0\right) \\ &P\left(W_{t_1} \le x_1, W_{t_2} \le x_2, \ldots, W_{t_k} \le x_k\right) \cdot P\left(W_1 = 0\right) \\ &P\left(W_{t_1} \le x_1, W_{t_2} \le x_2, \ldots, W_{t_k} \le x_k \middle| W(1) = 0\right) \\ &= \frac{P\left(W_{t_1} \le x_1, W_{t_2} \le x_2, \ldots, W_{t_k} \le x_k, W(1) = 0\right)}{P(W_1 = 0)} \\ &P\left(W_{t_1} \le x_1, W_{t_2} \le x_2, \ldots, W_{t_k} \le x_k\right). \end{aligned}$$

$\{W_t^0\}$ is called Brownian Bridge if

$$\begin{aligned} EW_t^0 W_s^0 &= E(W_t - tW(1))(W_s - sW(1)) \\ &= \min(t, s) - st - ts + ts \\ &= s(1-t) \end{aligned}$$

for $s \le t$.

The above calculations show the conditional distribution of Brownian Motion given $\{W(1) = 0\}$ is the distribution of $\{W_t^0, 0 \le t \le 1\}$.

2.10 Weak convergence of probability measure on Polish space

In chapter 3, we shall discuss the weak convergence in the space of functions with jump discontinuities with a topology that makes it a Polish space. This will be applied in chapter 4 for the weak convergence of semi-martingales. To make this book self-contained, we present to the readers the weak convergence of probability measures on Polish space.

Let (S, ρ) be a complete separable metric space. $\{P_n\}$, a sequence of probability measure on $\mathcal{B}(S)$, converges weakly to P if for all bounded continuous function on $\mathcal{X}$

$$\int f dP_n \longrightarrow \int f dP$$

and we write $P_n \Rightarrow P$. Let $\mathcal{B}(S) = \mathcal{S}$.

Theorem 2.5: Every probability measure P on $(S, \mathcal{S})$ is regular, that is, for every $\mathcal{S}$-set A and every ϵ there exist a closed set F and an open set G such that $F \subset A \subset G$ and $P(G - F) < \epsilon$.

Proof: Denote the metric on S by $\rho(x, y)$ and the distance from x to A by $\rho(x, A) = \inf\{\rho(x, y) : y \in A\}$. If A is closed, we can take $F = A$ and $G = A^\delta = \{x : \rho(x, A) < \delta\}$ for some δ, since the latter sets decrease to A as $\delta \downarrow 0$. Hence, we need only show that the class $\mathcal{G}$ of $\mathcal{S}$-sets with the asserted property is a σ-field. Given sets A_n in $\mathcal{G}$, choose closed sets F_n and open sets G_n such that $F_n \subset A_n \subset G_n$ and $P(G_n - F_n) < \epsilon/2^{n+1}$. If $G = \bigcup_n G_n$, and if $F = \bigcup_{n \le n_0} F_n$, with n_0 so chosen that

$$P\left(\bigcup_n F_n - F\right) < \frac{\epsilon}{2},$$

then $F \subset \bigcup_n A_n \subset G$ and $P(G - F) < \epsilon$. Thus, $\mathcal{G}$ is closed under the formation of countable unions; since it is obviously closed under complementation, $\mathcal{G}$ is a σ-field.

Equation (2.5) implies that P is completely determined by the values of $P(F)$ for closed sets F. The next theorem shows that P is also determined by the values of $\int f dP$ for bounded, continuous f. The proof depends on the approximation of indicator I_F by such an f, and the function $f(x) = (1 - \rho(x, F)/\epsilon)^+$ works. It is bounded, and it is continuous, even uniformly continuous, because $|f(x) - f(y)| \le \rho(x, y)/\epsilon$. $x \in F$ implies $f(x) = 1$, while $x \notin F^\epsilon$ implies $\rho(x, F) \ge \epsilon$ and hence $f(x) = 0$. Therefore,

$$I_F(x) \le f(x) = (1 - \rho(x, F)/\epsilon)^+ \le I_{F^\epsilon}(x). \tag{2.5}$$

Theorem 2.6: Probability measures P and Q on $\mathcal{S}$ coincide if and only if $\int f dP = \int f dQ$ for all bounded, uniformly continuous real functions f.

Proof: ($\Rightarrow$) Trivial.

($\Leftarrow$) For the bounded, uniformly continuous f of (2.5), $P(F) \le \int f dP = \int f dQ \le Q(F^\epsilon)$. Letting $\epsilon \downarrow 0$ gives $P(F) = Q(F)$, provided F is closed. By symmetry and (Theorem 2.5), $P = Q$.

The following notion of tightness plays a fundamental role both in the theory of weak convergence and in its applications. A probability measure P on $(S, \mathcal{S})$ is tight if for each ϵ there exists a compact set K such that $P(K) \ge 1 - \epsilon$. By (2.5), P is tight if and only if for each $A \in \mathcal{S}$

$$P(A) = \sup\{P(K) : K \subset A, \quad K \text{ is compact.}\}$$

Theorem 2.7: If S is separable and complete, then each probability measure on $(S, \mathcal{S})$ is tight.

Proof: Since S is separable, there is, for each k, a sequence $A_{k1}, A_{k2}, \dots$ of open $1/k$-balls covering S. Choose n_k large enough that

$$P\Big(\bigcup_{i \le n_k} A_{ki}\Big) > 1 - \frac{\epsilon}{2^k}.$$

By the completeness hypothesis, the totally bounded set

$$\bigcap_{k \ge 1} \bigcup_{i \le n_k} A_{ki}$$

has compact closure K. However, clearly, $P(K) > 1 - \epsilon$. This completes the proof.

The following theorem provides useful conditions equivalent to weak convergence; any of them could serve as the definition. A set A in $\mathcal{S}$ whose boundary ∂A satisfies $P(\partial A) = 0$ is called P-continuity set. Let P_n, P be probability measures on $(\mathcal{X}, \mathcal{B}(\mathcal{X}))$.

Theorem 2.8 (The Portmanteau theorem): The following are equivalent.

1. For bounded and continuous f

$$\lim_{n \to \infty} \int f dP_n = \int f dP.$$

2. For closed set F

$$\limsup_{n \to \infty} P_n(F) \le P(F).$$

3. For open set G

$$\liminf_{n \to \infty} P_n(G) \ge P(G).$$

4. For all set A with $P(\partial A) = 0$

$$\lim_{n\to\infty} P_n(A) = P(A).$$

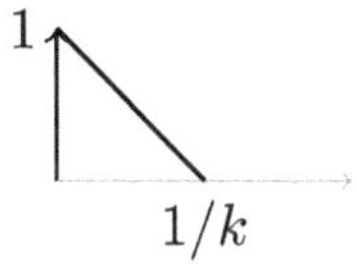

Proof: (1)→ (2) : Let $f_k(x) = \gamma_k(\rho(x, F))$, with $\gamma_k(x)$ given by the above graph. First of all, we know that

$$f_k(x) \searrow 1_F(x).$$

Then, for any $\delta > 0$, there exists K such that for all $k \geq K$

$$\begin{aligned}\limsup_{n\to\infty} P_n(F) &= \limsup_{n\to\infty} \int 1_F dP_n \\ &\leq \limsup_{n\to\infty} \int f_k dP_n \\ &= \lim_{n\to\infty} \int f_k dP_n \\ &= \int f_k dP \\ &\leq P(F) + \delta.\end{aligned}$$

The last inequality follows from the fact that

$$\int_F f_n dP \searrow P(F).$$

As a result, for all $\delta > 0$, we have

$$\limsup_{n\to\infty} P_n(F) \leq P(F) + \delta.$$

(2) → (3)

Let $G = F^c$. Then, it follows directly.

(2) + (3) → (4) trivial.

(4) → (1) Approximate f let simple functions f_n of sets A with $P(\bar{A}) = P(A)$.

Theorem 2.9: A necessary and sufficient condition for $P_n \Rightarrow P$ is that each subsequence $\{P_{n_i}\}$ contain a further subsequence $\{P_{n_i(m)}\}$ converging weakly to P.

Proof: The necessary is easy. As for sufficiency, if P_n does not converge weakly to P, then there exists some bounded and continuous f such that $\int f dP_n$ does not converge to $\int f dP$. However, for some positive ϵ and some subsequence P_{n_i},

$$\left| \int f dP_{n_i} - \int f dP \right| > \epsilon$$

for all i, and no further subsequence can converge weakly to P.

Suppose that h maps $\mathcal{X}$ into another metric space $\mathcal{X}'$, with metric ρ' and Borel σ–field $\mathcal{B}(\mathcal{X}')$. If h is measurable $\mathcal{X}/\mathcal{X}'$, then each probability P on $(\mathcal{X}, \mathcal{B}(\mathcal{X}))$ induces on $(\mathcal{X}', \mathcal{B}(\mathcal{X}'))$ a probability $P \circ h^{-1}$ defined as usual by $P \circ h^{-1}(A) = P(h^{-1}(A))$. We need conditions under which $P_n \Rightarrow P$ implies $P_n \circ h^{-1} \Rightarrow P \circ h^{-1}$. One such condition is that h is continuous: If f is bounded and continuous on $\mathcal{X}'$, then fh is bounded and continuous on $\mathcal{X}$, and by change of variable, $P_n \Rightarrow P$ implies

$$\int_{\mathcal{X}'} f(y) P_n \circ h^{-1}(dy) = \int_{\mathcal{X}} f(h(x)) P_n(dx) \to \int_{\mathcal{X}} f(h(x)) P(dx) = \int_{\mathcal{X}'} f(y) P \circ h^{-1}(dy) \tag{2.6}$$

Let D_h be discontinuity set of h.

Theorem 2.10: Let $(\mathcal{X}, \rho)$ and $(\mathcal{X}', \rho')$ be two Polish space and

$$h : \mathcal{X} \to \mathcal{X}'$$

with $P(D_h) = 0$. Then, $P_n \Rightarrow P$ implies

$$P_n \circ h^{-1} \Rightarrow P \circ h^{-1}.$$

Proof: Since

$$h^{-1}(F) \subset \overline{h^{-1}(F)} \subset D_h \cup h^{-1}(F),$$

$$\begin{aligned}
\limsup_{n \to \infty} P_n(h^{-1}(F)) &\le \limsup_{n \to \infty} P_n(\overline{h^{-1}(F)}) \\
&\le P\left(D_h \cup h^{-1}(F)\right) \\
&\le P(h^{-1}(F)) \text{ (since } D_h \text{ is a set of } P\text{-measure zero).}
\end{aligned}$$

Therefore, for all closed set F,

$$\limsup_{n \to \infty} P_n \circ h^{-1}(F) \le P \circ h^{-1}(F),$$

and hence, by Theorem 2.8, the proof is completed.

Let X_n and X be random variables($\mathcal{X}$-valued). Then, we say $X_n \to_{\mathcal{D}} X$ if $P \circ X_n^{-1} \Rightarrow P \circ X^{-1}$.

Observation: If $X_n \to_{\mathcal{D}} X$ and $\rho(X_n, Y_n) \to_p 0$, then

$$Y_n \to_{\mathcal{D}} X.$$

Remark: We use the following property of limsup and liminf.

$$\limsup_{n\to\infty}(a_n + b_n) \le \limsup_{n\to\infty} a_n + \limsup_{n\to\infty} b_n$$

and

$$\liminf_{n\to\infty}(a_n + b_n) \ge \liminf_{n\to\infty} a_n + \liminf_{n\to\infty} b_n.$$

Proof: Consider closed set F. Let $F^\epsilon = \{x : \rho(x, F) \le \epsilon\}$. Then, $F^\epsilon \searrow F$ as $\epsilon \to 0$ and

$$\begin{aligned}\{X_n \notin F^\epsilon\} &= \{\omega : X_n(\omega) \notin F^\epsilon\}\\ &= \{\omega : \rho(X_n(\omega), F) > \epsilon\}\end{aligned}$$

Therefore,

$$\begin{aligned}\omega \in \{X_n \notin F^\epsilon\} \cap \{\rho(X_n, Y_n) < \epsilon\} &\Rightarrow \rho(X_n(\omega), F) > \epsilon \text{ and } \rho(X_n(\omega), Y_n(\omega)) < \epsilon\\ &\Rightarrow \rho(Y_n(\omega), F) > 0 \text{ (draw graph.)}\\ &\Rightarrow Y_n(\omega) \notin F\\ &\Rightarrow \omega \in \{Y_n \notin F\}\end{aligned}$$

Thus,

$$\{X_n \notin F^\epsilon\} \cap \{\rho(X_n, Y_n) < \epsilon\} \subset \{Y_n \notin F\}.$$

Therefore,

$$P(Y_n \in F) \le P(\rho(X_n, Y_n) > \epsilon) + P(X_n \in F^\epsilon).$$

Let $P_n^Y = P \circ Y_n^{-1}$ and $P^X = P \circ X^{-1}$. Then, for all $\epsilon > 0$,

$$\begin{aligned}\limsup_{n\to\infty} P_n(F) &= \limsup_{n\to\infty} P(Y_n \in F)\\ &\le \limsup_{n\to\infty} P(\underbrace{\rho(X_n, Y_n)}_{\to_p 0} > \epsilon) + \limsup_{n\to\infty} P(X_n \in F^\epsilon)\end{aligned}$$

$$= \limsup_{n \to \infty} P(X_n \in F^\epsilon)$$
$$= P(X \in F^\epsilon) \text{ (since } X_n \Rightarrow_{\mathcal{D}} X).$$

Therefore, for all closed set F, we have

$$\limsup_{n \to \infty} P_n^Y(F) \le P^X(F),$$

and hence, by Theorem 2.5,

$$P_n^Y \Longrightarrow P^X,$$

which implies $Y_n \Rightarrow_{\mathcal{D}} X$.

We say that a family of probability measure $\Pi \subset \mathcal{P}(\mathcal{X})$ is tight if, given $\epsilon > 0$, there exists compact K_ϵ such that

$$P(K_\epsilon) > 1 - \epsilon \text{ for all } P \in \Pi.$$

2.10.1 Prokhorov theorem

Definition 2.4: Π is relatively compact if, for $\{P_n\} \subset \Pi$, there exists a subsequence $\{P_{n_i}\} \subset \Pi$ and probability measure P (not necessarily an element of Π), such that

$$P_{n_i} \Rightarrow P.$$

Even though $P_{n_i} \Rightarrow P$ makes no sense if $P(\mathcal{X}) < 1$, it is to be emphasized that we do require $P(\mathcal{X}) = 1$ and we disallow any escape of mass, as discussed below. For the most part, we are concerned with the relative compactness of sequences $\{P_n\}$; this means that every subsequence $\{P_{n_i}\}$ contains a further subsequence $\{P_{n_i(m)}\}$, such that $P_{n_i(m)} \Rightarrow P$ for some probability measure P.

Example: Suppose we know of probability measures P_n and P on $(C, \mathcal{C})$ that the finite-dimensional distributions of P_n converges weakly to those of P: $P_n\pi^{-1}_{t_1,\dots,t_k} \Rightarrow P\pi^{-1}_{t_1,\dots,t_k}$ for all k and all $t_1, \dots, t_k$. Notice that P_n need not converge weakly to P. Suppose, however, that we also know that $\{P_n\}$ is relatively compact. Then each $\{P_n\}$ contains some $\{P_{n_i(m)}\}$ converging weakly to some Q. Since the mapping theorem then gives $P_{n_i(m)}\pi^{-1}_{t_1,\dots,t_k} \Rightarrow Q\pi^{-1}_{t_1,\dots,t_k}$ and since $P_n\pi^{-1}_{t_1,\dots,t_k} \Rightarrow P\pi^{-1}_{t_1,\dots,t_k}$ by assumption, we have $P\pi^{-1}_{t_1,\dots,t_k} = Q\pi^{-1}_{t_1,\dots,t_k}$ for all $t_1, \dots, t_k$. Thus, the finite-dimensional distributions of P and Q are identical, and since the class $\mathcal{C}_f$ of finite-dimensional sets is a separating class, $P = Q$. Therefore, each subsequence contains a further subsequence converging weakly to P, not to some fortuitous limit, but specifically to P. It follows by Theorem 2.9 that the entire sequence $\{P_n\}$ converges weakly to P.

Therefore, $\{P_n\}$ is relatively compact and the finite-dimensional distributions of P_n converge weakly to those of P, then $P_n \Rightarrow P$. This idea provides a powerful method for proving weak convergence in C and other function spaces. Note that if $\{P_n\}$ does converge weakly to P, then it is relatively compact, so that this is not too strong a condition.

Theorem 2.11: Suppose $(\mathcal{X}, \rho)$ is a Polish space and $\Pi \subset \mathcal{P}(\mathcal{X})$ is relatively compact, then it is tight.

This is the converse half of Prohorov's theorem.

Proof: Consider open sets, $G_n \nearrow \mathcal{X}$. For each $\epsilon > 0$, there exists n, such that for all $P \in \Pi$

$$P(G_n) > 1 - \epsilon.$$

Otherwise, for each n, we can find P_n such that $P_n(G_n) < 1 - \epsilon$. Then by by relative compactness, there exists $\{P_{n_i}\} \subset \Pi$ and probability measure $Q \in \Pi$ such that $P_{n_i} \Rightarrow Q$. Thus,

$$\begin{aligned} Q(G_n) &\leq \liminf_{i\to\infty} P_{n_i}(G_n) \\ &\leq \liminf_{i\to\infty} P_{n_i}(G_{n_i}) \text{ (since } n_i \geq n \text{ and hence } G_n \subset G_{n_i}) \\ &< 1 - \epsilon \end{aligned}$$

Since $G_n \nearrow \mathcal{X}$,

$$\begin{aligned} 1 &= Q(\mathcal{X}) \\ &= \lim_{n\to\infty} Q(G_n) \\ &< 1 - \epsilon, \end{aligned}$$

which is contradiction. Let A_{k_m}, $m = 1, 2, \dots$ be open ball with radius $\frac{1}{k_m}$, covering $\mathcal{X}$(separability). Then, there exists n_k such that for all $P \in \Pi$,

$$P\Big(\bigcup_{i\leq n_k} A_{k_i}\Big) > 1 - \frac{\epsilon}{2^k}$$

Then, let

$$K_\epsilon = \overline{\bigcap_{k\geq 1} \bigcup_{i\leq n_k} A_{k_i}},$$

where $\bigcap_{k\geq 1}\bigcup_{i\leq n_k} A_{k_i}$ is totally bounded set. Then, K_ϵ is compact(completeness), and $P(K_\epsilon) > 1 - \epsilon$.

Remark: The last inequality is from the following. Let B_i be such that $P(B_i) > 1 - \frac{\epsilon}{2^i}$. Then,

$$\begin{aligned} P(B_i) > 1 - \frac{\epsilon}{2^i} &\Rightarrow P(B_i^c) \leq \frac{\epsilon}{2^i} \\ &\Rightarrow P(\cup_{i=1}^{\infty} B_i^c) \leq \epsilon \\ &\Rightarrow P(\cap_{i=1}^{\infty} B_i) > 1 - \epsilon. \end{aligned}$$

2.11 Tightness and compactness in weak convergence

Theorem 2.12: If Π is tight, then for $\{P_n\} \subset \Pi$, there exists a subsequence $\{P_{n_i}\} \subset \{P_n\}$ and probability measure P such that

$$P_{n_i} \Rightarrow P.$$

Proof: Choose compact $K_1 \subset K_2 \subset \ldots$, such that for all n

$$P_n(K_u) > 1 - \frac{1}{u}$$

from tightness condition. Look at $\bigcup_u K_u$. We know that there exists a countable family of open sets, $\mathcal{A}$, such that if $x \in \cup_u K_u$ and G is open, then

$$x \in A \subset \bar{A} \subset G$$

for some $A \in \mathcal{A}$. Let

$$\mathcal{H} = \{\emptyset\} \cup \{\text{finite union of sets of the form } \bar{A} \cap K_u\ u \geq 1, A \in \mathcal{A}\}.$$

Then, $\mathcal{H}$ is a countable family. Using the Cantor diagonalization method, there exists $\{n_i\}$ such that for all $H \in \mathcal{H}$,

$$\alpha(H) = \lim_{i\to\infty} P_{n_i}(H).$$

Our aim is to construct a probability measure P such that for all open set G,

$$P(G) = \sup_{H\subset G} \alpha(H). \tag{2.7}$$

Suppose we showed (2.7) above. Consider an open set G. Then, for $\epsilon > 0$, there exists $H_\epsilon \subset G$ such that

$$\begin{aligned}
P(G) &= \sup_{H \subset G} \alpha(H) \\
&< \alpha(H_\epsilon) + \epsilon \\
&= \lim_i P_{n_i}(H_\epsilon) + \epsilon \\
&= \liminf_i P_{n_i}(H_\epsilon) + \epsilon \\
&\leq \liminf_i P_{n_i}(G) + \epsilon
\end{aligned}$$

and hence, for all open set G,

$$P(G) \leq \liminf_i P_{n_i}(G),$$

which is equivalent to $P_{n_i} \Rightarrow P$.

Observe $\mathcal{H}$ is closed under finite union and

1. $\alpha(H_1) \leq \alpha(H_2)$ if $H_1 \subset H_2$
2. $\alpha(H_1 \cup H_2) = \alpha(H_1) + \alpha(H_2)$ if $H_1 \cap H_2 = \emptyset$
3. $\alpha(H_1 \cup H_2) \leq \alpha(H_1) + \alpha(H_2)$.

Define for open set G

$$\beta(G) = \sup_{H \subset G} \alpha(H). \tag{2.8}$$

Then, $\alpha(\emptyset) = \beta(\emptyset) = 0$ and β is monotone.

Define for $M \subset \mathcal{X}$

$$\gamma(M) = \inf_{M \subset G} \beta(G).$$

Then,

$$\begin{aligned}
\gamma(M) &= \inf_{M \subset G} \beta(G) \\
&= \inf_{M \subset G} \big(\sup_{H \subset G} \alpha(H) \big) \\
\gamma(G) &= \inf_{G \subset G'} \beta(G') \\
&= \beta(G)
\end{aligned}$$

M is γ-measurable if for all $L \subset \mathcal{X}$

$$\gamma(L) \geq \gamma(M \cap L) + \gamma(M^c \cap L).$$

We shall prove that γ is outer measure, and hence open and closed sets are β-measurable.

γ-measurable sets M form a σ-field, $\mathcal{M}$, and

$$\gamma\Big|_{\mathcal{M}}$$

is a measure.

Claim: Each closed set is in $\mathcal{M}$ and

$$P = \gamma\Big|_{\mathcal{B}(\mathcal{X})}$$

open set G

$$P(G) = \gamma(G) = \beta(G).$$

Note that P is a probability measure. K_u has finite covering of sets in $\mathcal{A}$ when $K_u \in \mathcal{H}$.

$$\begin{aligned} 1 &\geq P(\mathcal{X}) \\ &= \beta(\mathcal{X}) \\ &= \sup_u \alpha(K_u) \\ &= \sup_u \left(1 - \frac{1}{u}\right) \\ &= 1. \end{aligned}$$

Step 1: If $F \subset G$ (F is closed and G is open), and if $F \subset H$ for some $H \in \mathcal{H}$, then there exists some $H_0 \in \mathcal{H}$ such that

$$H \subset H_0 \subset G.$$

Proof: Consider $x \in F$ and $A_x \in \mathcal{A}$ such that

$$x \in A_x \subset \bar{A}_x \subset G.$$

Since F is closed subset of compact, F is compact. Since A_x covers F, there exists finite subcovers $A_{x_1}, A_{x_2}, \ldots, A_{x_k}$. Take

$$H_0 = \bigcup_{i=1}^{k} (\bar{A}_{x_i} \cap K_u).$$

Step 2: β is finitely sub-additive on open set. Suppose $H \subset G_1 \cup G_2$ and $H \in \mathcal{H}$. Let

$$F_1 = \{x \in H : \rho(x, G_1^c) \geq \rho(x, G_2^c)\}$$

$$F_2 = \{x \in H : \rho(x, G_2^c) \geq \rho(x, G_1^c)\}.$$

If $x \in F_1$ but not in G_1, then $x \in G_2$, and hence $x \in H$. Suppose x is not in G_2. Then $x \in G_2^c$, and hence, $\rho(x, G_2^c) > 0$. Therefore,

$$\begin{aligned} 0 &= \rho(x, G_1^c) \text{ (since } x \in G_1^c) \\ &< \underbrace{\rho(x, G_2^c)}_{>0}, \end{aligned}$$

which contradicts $x \in F_1$, and hence contradicts $\rho(x, G_1^c) \geq \rho(x, G_2^c)$. Similarly, if $x \in F_2$ but not in G_2, then $x \in G_1$. Therefore, $F_1 \subset G_1$ and $F_2 \subset G_2$. Since F_i's are closed, by step 1, there exist H_1 and H_2 such that

$$F_1 \subset H_1 \subset G_1$$

and

$$F_2 \subset H_2 \subset G_2.$$

Therefore,

$$\begin{aligned} \alpha(H) &\leq \alpha(H_1) + \alpha(H_2) \\ \beta(G) &\leq \beta(G_1) + \beta(G_2). \end{aligned}$$

Step 3: β is countably sub-additive on open-set $H \subset \bigcup_n G_n$, where G_n is an open set. Since H is compact (union of compacts), there exist a finite subcovers, i.e., there exists n_0 such that

$$H \subset \bigcup_{n \leq n_0} G_n$$

and

$$\begin{aligned} \alpha(H) &\leq \beta(H) \\ &\leq \beta\Big(\bigcup_{n \leq n_0} G_n\Big) \\ &= \sum_{n \leq n_0} \beta(G_n) \\ &= \sum_n \beta(G_n). \end{aligned}$$

Therefore,

$$\begin{aligned} \beta\Big(\bigcup_n G_n\Big) &= \sup_{H \subset \bigcup_n G_n} \alpha(H) \\ &\leq \sup_{H \subset \bigcup_n G_n} \sum_n \beta(G_n) \\ &= \sum_n \beta(G_n). \end{aligned}$$

Step 4: γ is an outer measure. We know γ is monotonic by definition and is countably sub-additive. Given $\epsilon > 0$ and subsets $\{M_n\} \subset \mathcal{X}$, choose open sets G_n, $M_n \subset G_n$ such that

$$\beta(G_n) \leq \gamma(M_n) + \frac{\epsilon}{2^n}$$

$$\begin{aligned}\gamma\Big(\bigcup_n M_n\Big) &\leq \beta\Big(\bigcup_n G_n\Big)\\ &= \sum \beta(G_n)\\ &= \sum_n \gamma(M_n) + \epsilon.\end{aligned}$$

Step 5: F is closed G is open.

$$\beta(G) \geq \gamma(F \cap G) + \gamma(F^c \cap G).$$

Choose $\epsilon > 0$ and $H_1 \in \mathcal{H}$, $H_1 \subset F^c \cap G$ such that

$$\alpha(H_1) > \beta(G \cap F^c) - \epsilon.$$

Chose H_0 such that

$$\alpha(H_0) > \beta(H_1^c \cap G) - \epsilon.$$

Then, $H_0, H_1 \subset G$, and $H_0 \cap H_1 = \emptyset$,

$$\begin{aligned}\beta(G) &\geq \alpha(H_0 \cup H_1)\\ &= \alpha(H_0) + \alpha(H_1)\\ &> \beta(H_1^c \cap G) + \beta(F^c \cap G) - 2\epsilon\\ &\geq \gamma(F \cap G) + \gamma(F^c \cap G) - 2\epsilon.\end{aligned}$$

Step 6: If $F \in \mathcal{M}$, then F are all closed. If G is open and $L \subset G$, then

$$\beta(G) \geq \gamma(F \cap L) + \gamma(F^c \cap L).$$

Then,

$$\begin{aligned}&\inf \beta(G) \geq \inf\left(\gamma(F \cap L) + \gamma(F^c \cap L)\right)\\ \Longrightarrow\ &\gamma(L) \geq \gamma(F \cap L) + \gamma(F^c \cap L).\end{aligned}$$

3 Weak convergence on $C[0, 1]$ and $D[0, \infty)$

We now use the techniques developed in sections 2.10 and 2.11 to the case of space $C[0, 1]$, which is a complete separable metric space with sup norm. In later sections, we consider the space of functions with discontinuities of the first kind. Clearly, one has to define distance on this space denoted by $D[0, 1]$, such that the space is complete separable metric space. This was done by Skorokhod with the introduction of the topology called Skorokhod topology. In view of the fact that convergence of finite dimensional distributions determines the limiting measure and the Prokhorov theorem from chapter 2, we need to characterize compact sets in $C[0, 1]$ with sup norm and $D[0, 1]$ with Skorokhod topology, which is described in section 3.4. The tightness in this case is described in section 3.7.

3.1 Structure of compact sets in $C[0, 1]$

Let $\mathcal{X}$ be a complete separable metric space. We showed that Π is tight if and only if Π is relatively compact. Consider P_n is measure on $C[0, 1]$ and let

$$\pi_{t_1,\ldots,t_k}(x) = (x(t_1), \ldots, x(t_k))$$

and suppose that

$$P_n \circ \pi^{-1}_{t_1,\ldots,t_k} \Longrightarrow P \circ \pi^{-1}_{t_1,\ldots,t_k}$$

does not imply

$$P_n \Longrightarrow P$$

on $C[0, 1]$. However, $P_n \circ \pi^{-1}_{t_1,\ldots,t_k} \Rightarrow P \circ \pi^{-1}_{t_1,\ldots,t_k}$ and $\{P_n\}$ is tight. Then, $P_n \Rightarrow P$.

Proof: Since tightness is implied (as we proved), there exists a subsequence $\{P_{n_i}\}$ of $\{P_n\}$ such that the sequence converges weakly to a probability measure Q, i.e $P_{n_i} \Longrightarrow Q$.

Hence, by Theorem 2.9, $P_{ni} \circ \pi^{-1}_{t_1 \cdots t_k} \longrightarrow Q \circ \pi^{-1}_{t_1 \ldots t_k}$, Giving $P = Q$.

However, all subsequences have the same limit. Hence, $P_n(\rightarrow)P$.

What is a compact set in $C[0, 1]$?

3.1.1 Arzela-Ascoli theorem

Definition 3.1: The uniform norm (or sup norm) assigns to real- or complex-valued bounded functions f defined on a set S the non-negative number

$$||f||_\infty = ||f||_{\infty,S} = \sup\{|f(x)| : x \in S\}.$$

This norm is also called the supremum norm, the Chebyshev norm, or the infinity norm. The name "uniform norm" derives from the fact that a sequence of functions $\{f_n\}$ converges to f under the metric derived from the uniform norm if and only if f_n converges to f uniformly.

Theorem 3.1: The set $A \subset C[0,1]$ is relatively compact in sup topology if and only if

$$\text{(i)}\ \sup_{x\in A} |x(0)| < \infty.$$

$$\text{(ii)}\ \lim_{\delta} \left(\sup_{x\in A} w_x(\delta) \right) = 0.$$

Remark (modulus of continuity): Here

$$w_x(\delta) = \sup_{|s-t|\le\delta} |x(t) - x(s)|.$$

Proof: Consider function

$$f : C[0,1] \to R,$$

such that $f(x) = x(0)$.

Claim: f is continuous.

Proof of claim: We want to show that for $\epsilon > 0$, there exists δ such that

$$||x-y||_\infty = \sup_{t\in[0,1]} \{|x(t)-y(t)|\} < \delta \text{ implies } |f(x)-f(y)| = |x(0)-y(0)| < \delta.$$

Given $\epsilon > 0$, let $\delta = \epsilon$. Then, we are done.

Since A is compact, continuous mapping $x \mapsto x(0)$ is bounded. Therefore,

$$\sup_{x\in A} |x(0)| < \infty.$$

$w_x\left(\frac{1}{n}\right)$ is continuous in x uniformly on A and hence

$$\lim_{n\to\infty} w_x\left(\frac{1}{n}\right) = 0.$$

Suppose (i) and (ii) hold. Choose k large enough so that

$$\sup_{x\in A} w_x\left(\frac{1}{k}\right) = \sup_{x\in A} \Big(\sup_{|s-t|\le\frac{1}{k}} |x(s)-x(t)| \Big)$$

is finite. Since

$$|x(t)| < |x(0)| + \sum_{i=1}^{k} \left|x(\frac{it}{k}) - x(\frac{(i-1)t}{k})\right|,$$

we have

$$\alpha = \sup_{0\le t\le 1} \left(\sup_{x\in A} |x(t)| \right) < \infty.$$

Choose $\epsilon > 0$ and finite ϵ-covering H of $[-\alpha, \alpha]$. Choose k large enough so that

$$w_x\left(\frac{1}{k}\right) < \epsilon.$$

Take B to be finite set of polygonal functions on $C[0,1]$ that are linear on $\left[\frac{i-1}{k}, \frac{i}{k}\right]$ and takes the values in H at end points.

If $x \in A$ and $\left|x(\frac{1}{k})\right| \le \alpha$ so that there exists a point $y \in B$ such that

$$\left|x\left(\frac{i}{k}\right) - y\left(\frac{i}{k}\right)\right| < \epsilon, \quad i = 1, 2, \ldots, k$$

then

$$\left|y\left(\frac{i}{k}\right) - x(t)\right| < 2\epsilon \quad \text{for } t \in \left[\frac{i-1}{k}, \frac{i}{k}\right].$$

$y(t)$ is convex combination of $y(\frac{i}{k}), y(\frac{i-1}{k})$, so it is within 2ϵ of $x(t)$, $||x - y||_\infty < 2\epsilon$, B is finite, B is 2ϵ-covering of A. This implies A is compact.

Theorem 3.2: $\{P_n\}$ is tight on $C[0,1]$ if and only if

1. For each $\eta > 0$, there exists a and n_0 such that for $n \ge n_0$

$$P_n(\{x : |x(0)| > a\}) < \eta$$

2. For each $\epsilon, \eta > 0$, there exists $0 < \delta < 1$ and n_0 such that for $n \ge n_0$

$$P_n(\{x : w_x(\delta) \ge \epsilon\}) < \eta$$

Proof: Since $\{P_n\}$ is tight, given $\delta > 0$, choose K compact such that

$$P_n(K) > 1 - \eta.$$

Note that by the Arzela-Ascoli theorem, for large a

$$K \subset \{x : |x(0)| \leq a\}$$

and for small δ

$$K \subset \{x : w_x(\delta) \leq \epsilon\}.$$

Now, $C[0,1]$ is a complete separable metric space. So for each n, P_n is tight, and hence, we get the necessity condition. Given $\eta > 0$, there exists a such that

$$P(\{x : |x(0)| > a\}) < \eta$$

and $\epsilon, \eta > 0$, there exists $\delta > 0$ such that

$$P_n(\{x : w_x(\delta) \geq \epsilon\}) < \eta.$$

This happens for P_n, where $n \leq n_0$ with n_0 is finite. Assume (i) and (ii) holds for all n. Given η, choose a so that

$$B = \{x : |x(0)| \leq a\},$$

a satisfies for all n

$$P_n(B) > 1 - \eta,$$

and choose δ_k such that

$$B_k = \left\{x : w_x(\delta_k) \leq \frac{1}{k}\right\},$$

with

$$P_n(B_k) > 1 - \frac{\eta}{2^k}.$$

Let $K = \overline{A}$ where

$$A = B \cap \left(\bigcap_k B_k\right).$$

K is compact by Arzela-Ascoli Theorem,

$$P_n(K) > 1 - 2\eta.$$

3.2 Invariance principle of sums of i.i.d. random variables

Let X_i's be i.i.d. with $EX_i = 0$ and $EX_i^2 = \sigma^2$. Define

$$W_t^n(\omega) = \frac{1}{\sigma\sqrt{n}} S_{[nt]}(\omega) + (nt - [nt]) \frac{1}{\sigma^2\sqrt{n}} X_{[nt]+1}.$$

Consider linear interpolation

$$W_{\frac{k}{n}} = \frac{S_k}{\sqrt{n}} \quad \text{where } W^n \in C[0,1] \text{ a.e. } P_n.$$

Let

$$\psi_{nt} = (nt - [nt]) \cdot \frac{X_{[nt+1]}}{\sigma\sqrt{n}}.$$

Claim: For fixed t, by Chevyshev's inequality, as $n \to \infty$

$$\psi_{nt} \to 0$$

Proof of Claim:

$$\begin{aligned} P(|\psi_{nt}|\epsilon) &= P\left(|X_{[nt+1]}| > \frac{\sigma\sqrt{n}\epsilon}{(nt - [nt])} \right) \\ &\le \frac{E|X_{[nt+1]}|^2}{\frac{\sigma^2 n\epsilon^2}{(nt-[nt])^2}} \\ &= \frac{(nt - [nt])^2}{n\epsilon^2} \\ &\le \frac{1}{n\epsilon^2} \to 0. \end{aligned} \tag{3.1}$$

By CLT,

$$\frac{S_{[nt]}}{\sigma\sqrt{[nt]}} \Rightarrow_{\mathcal{D}} N(0,1).$$

Since $\frac{[nt]}{n} \to t$, by CLT,

$$\frac{S_{[nt]}}{\sigma\sqrt{n}} = \frac{S_{[nt]}}{\sigma\sqrt{[nt]}} \times \frac{\sqrt{[nt]}}{\sqrt{n}} \Rightarrow_{\mathcal{D}} \sqrt{t}Z$$

by Slutsky's equation. Therefore,

$$W_t^n \Rightarrow_{\mathcal{D}} \sqrt{t}Z.$$

Then,

$$\begin{aligned}(W_s^n, W_t^n - W_s^n) &= \frac{1}{\sigma\sqrt{n}}\left(S_{[ns]}, S_{[nt]} - S_{[ns]}\right) + \left(\psi_{ns}, \psi_{nt} - \psi_{ns}\right) \\ &\Longrightarrow_{\mathcal{D}} (N_1, N_2).\end{aligned}$$

Since $S_{[ns]}$ and $S_{[nt]} - S_{[ns]}$ are independent, N_1 and N_2 are independent normal with variance s and $t - s$. Thus,

$$\begin{aligned}(W_s^n, W_t^n) &= (W_s^n, (W_t^n - W_s^n) + W_s^n) \\ &\Rightarrow_{\mathcal{D}} (N_1, N_1 + N_2).\end{aligned}$$

The two-dimensional distributions of $(W_t^n)_{t\in[0,1]}$ converges to two dimensional distributions of Brownian Motion. We consider similarly k-dimensional distributions of $(W_t^n)_{t\in[0,1]}$ converge to those of Brownian Motion.

We considered two-dimensional. We can take k–dimensional. Similar argument shows that

$$(W_1^n, \ldots, W_k^n) \Rightarrow_{\mathcal{D}} \text{ finite dimensional distribution of Brownian motion.}$$

Now, we have to show that P_n is tight. Recall the Arzela-Ascoli theorem.

Theorem 3.3: $\{P_n\}$ is tight on $C[0,1]$ if and only if

1. For each $\eta > 0$, there exists a and n_0 such that for $n \geq n_0$

$$P_n(\{x : |x(0)| > a\}) > \eta.$$

2. For each $\epsilon, \eta > 0$, there exists $0 < \delta < 1$ and n_0 such that for $n \geq n_0$

$$P_n(\{x : w_x(\delta) \geq \epsilon\}) < \eta.$$

Theorem 3.4: Suppose $0 = t_0 < t_1 < \cdots < t_\nu = 1$ and

$$\min_{1<i<\nu}(t_i - t_{i-1}) \geq \delta. \tag{3.2}$$

Then for arbitrary x,

$$w_x(\delta) \leq 3\max_{1\leq i\leq \nu}\left(\sup_{t_{i-1}\leq s\leq t_i} |x(s) - x(t_{i-1})|\right) \tag{3.3}$$

and for any P on $C[0,1]$

$$P(x : w_x(\delta) \geq 3\epsilon) \leq \sum_{i=1}^{\nu} P(x : \sup_{t_{i-1}\leq s\leq t_i} |x(s) - x(t_{i-1})| \geq \epsilon). \tag{3.4}$$

Proof: Let m denote the maximum in (3.3), i.e.

$$m = \max_{1\le i\le \nu}\Big(\sup_{t_{i-1}\le s\le t_i}|x(s)-x(t_{i-1})|\Big).$$

If s, t lie in $I_i = [t_{i-1}, t_i]$. Then

$$\begin{aligned}|x(s)-x(t)| &\le |x(s)-x(t_{i-1})| + |x(t)-x(t_{i-1})|\\ &\le 2m.\end{aligned}$$

Suppose s, t lie in adjoining intervals I_{i-1} and I_i. Then,

$$\begin{aligned}|x(s)-x(t)| &\le |x(s)-x(t_{i-1})| + |x(t_i)-x(t_{i-1})| + |x(t)-x(t_i)|\\ &\le 3m.\end{aligned}$$

Since

$$\min_{1<i<\nu}(t_i - t_{i-1}) \ge \delta.$$

for s and t to be such that $|s - t| < \delta$, s and t should lie in the same interval or adjoining intervals. Therefore,

$$\begin{aligned}w_x(\delta) &= \sup_{|s-t|\le\delta}|x(t)-x(s)|\\ &\le \max\left\{\sup_{s,t\in\text{same interval}}|x(t)-x(s)|,\ \sup_{s,t\in\text{adjoining interval}}|x(t)-x(s)|\right\}\\ &\le 3m.\end{aligned}$$

This proves (3.3). Note that if $X \ge Y$, then

$$P(X > a) \ge P(Y > a).$$

Therefore,

$$\begin{aligned}P(x : w_x(\delta) > 3\epsilon) &\le P\Big(3\max_{1\le i\le\nu}\Big(\sup_{t_{i-1}\le s\le t_i}|x(s)-x(t_{i-1})|\Big) > 3\epsilon\Big)\\ &= P\left(x : \max_{1\le i\le\nu}\Big(\sup_{t_{i-1}\le s\le t_i}|x(s)-x(t_{i-1})|\Big) > \epsilon\right)\\ &= \sum_{i=1}^{\nu} P\Big(x : \sup_{t_{i-1}\le s\le t_i}|x(s)-x(t_{i-1})| > \epsilon\Big).\end{aligned}$$

This proves the theorem.

Condition (ii) of the Arzela-Ascoli theorem holds if for each ϵ, η, there exists $\delta \in (0, 1)$ and n_0 such that for all $n \ge n_0$

$$\frac{1}{\delta}P_n\left(x : \sup_{t\le s\le t+\delta}|x(s)-x(t)| \ge \epsilon\right) > \eta.$$

Now apply Theorem 3.4 with $t_i = i\delta$ for $i < \nu = [1/\delta]$. Then by (3.4), condition (ii) of Theorem 3.3 holds.

3.3 Invariance principle for sums of stationary sequences

Definition 3.2: $\{X_n\}$ is stationary if for any m,

$$(X_{i_i}, \dots, X_{i_k}) =^{\mathcal{D}} (X_{i_i+m}, \dots, X_{i_k+m}).$$

Lemma 3.1: Suppose $\{X_n\}$ is stationary and W^n is defined as above. If

$$\lim_{\lambda\to\infty} \limsup_{n\to\infty} \lambda^2 P\Big(\max_{k\le n} |S_k| > \lambda\sigma\sqrt{n}\Big) = 0$$

then, W^n is tight.

Proof: Since $W_0^n = 0$, the condition (i) of Theorem 3.3 is satisfied. Let P_n is induced measure of W^n, i.e. consider now $P_n(w(\delta) > \epsilon)$, we shall show that for all $\epsilon > 0$

$$\lim_{\delta\to 0} \limsup_{n\to\infty} P_n\Big(w^{(\delta)} \ge \epsilon\Big) = 0.$$

If

$$\min(t_t - t_{i-1}) \ge \delta,$$

then by Theorem 3.4,

$$P\Big(w(W^n, \delta) \ge 3\epsilon\Big) \le \sum_{i=1}^{\nu} P\Big(\sup_{t_{i-1}\le s\le t_i} |W_s^n - W_t^n| \ge \epsilon\Big).$$

Take $t_i = \frac{m_i}{n}$, $0 = m_0 < m_1 < \cdots < m_\nu = n$. W_t^n is polygonal and hence,

$$\sup_{t_{i-1}\le s\le t_i} |W_s^n - W_t^n| = \max_{m_{i-1}\le k\le m_i} \frac{|S_k - S_{m_{i-1}}|}{\sigma\sqrt{n}}.$$

Therefore,

$$\begin{aligned} P\Big(w(W^n, \delta) \ge 3\epsilon\Big) &\le \sum_{i=1}^{\nu} P\Big(\max_{m_{i-1}\le k\le m_i} \frac{|S_k - S_{m_{i-1}}|}{\sigma\sqrt{n}} \ge \epsilon\Big) \\ &\le \sum_{i=1}^{\nu} P\Big(\max_{m_{i-1}\le k\le m_i} |S_k - S_{m_{i-1}}| \ge \sigma\sqrt{n}\epsilon\Big) \\ &= \sum_{i=1}^{\nu} P\Big(\max_{k\le m_i - m_{i-1}} |S_k| \ge \sigma\sqrt{n}\epsilon\Big) \text{ (by stationarity).} \end{aligned}$$

This inequality holds if

$$\frac{m_i}{n} - \frac{m_{i-1}}{n} \geq \delta \text{ for } 1 < i < \nu.$$

Take $m_i = im$ for $0 \leq i < \nu$ and $m_\nu = n$. For $i < \nu$ choose δ such that

$$m_i - m_{i-1} = m \geq n\delta.$$

Let $m = [n\delta]$, $\nu = \left[\frac{n}{m}\right]$. Then,

$$m_\nu - m_{\nu-1} \leq m$$

and

$$\nu = \left[\frac{n}{m}\right] \longrightarrow \frac{1}{\delta} \text{ where } \frac{1}{2\delta} < \frac{1}{\delta} < \frac{2}{\delta}.$$

Therefore, for large n

$$\begin{aligned} P\Big(w(W^n,\delta) \geq 3\epsilon\Big) &\leq \sum_{i=1}^{\nu} P\Big(\max_{k \leq m_i - m_{i-1}} |S_k| \geq \sigma\sqrt{n}\epsilon\Big) \\ &\leq \nu P\Big(\max_{k\leq m} |S_k| \geq \sigma\sqrt{n}\epsilon\Big) \\ &\leq \frac{2}{\delta} P\Big(\max_{k\leq m} |S_k| \geq \sigma\sqrt{n}\epsilon\Big). \end{aligned}$$

Take $\lambda = \frac{\epsilon}{\sqrt{2\delta}}$. Then,

$$P\Big(w(W^n,\delta) \geq 3\epsilon\Big) \leq \frac{4\lambda^2}{\epsilon^2} P\Big(\max_{k\leq m} |S_k| \geq \lambda\sigma\sqrt{n}\Big).$$

By the condition of the Lemma, given $\epsilon, \eta > 0$, there exists $\lambda > 0$ such that

$$\frac{4\lambda^2}{\epsilon^2} \limsup_{n\to\infty} P\Big(\max_{k\leq n} |S_k| > \lambda\sigma\sqrt{n}\Big) < \eta.$$

Now, for fixed λ, δ, let $m \to \infty$ with $n \to \infty$.

Look at X_k i.i.d. Then,

$$\lim_{\lambda\to\infty} \limsup_{n\to\infty} P\Big(\max_{k\leq n} |S_k| > \lambda\sigma\sqrt{n}\Big) = 0.$$

We know that

$$P\Big(\max_{u\leq m} |S_u| > \alpha\Big) \leq 3 \max_{u\leq m} P\Big(|S_u| > \frac{\alpha}{3}\Big).$$

To show

$$\lim_{\lambda\to\infty}\limsup_{n\to\infty}\lambda^2 P\Big(\max_{k\le n}|S_k| > \lambda\sigma\sqrt{n}\Big) = 0, \qquad (A),$$

we assume that X_i i.i.d. normal, and hence, $S_k/\sqrt{k}$ is asymptotically normal, N. Since we know

$$P(|N| > \lambda) \le \frac{EN^4}{\lambda^4} = \frac{3\sigma^4}{\lambda^4},$$

we have for $k \le n$, $\left(\frac{\sqrt{n}}{\sqrt{k}} > 1\right)$

$$\begin{aligned} P(|S_k| > \lambda\sigma\sqrt{n}) &= P(\sqrt{k}|N| > \lambda\sigma\sqrt{n}) \\ &\le \frac{3}{\lambda^4 T^4}. \end{aligned}$$

k_λ is large and $k_\lambda \le k \le n$. Then,

$$\begin{aligned} P(|S_k| > \lambda\sigma\sqrt{n}) &\le P(|S_k| > \lambda\sigma\sqrt{k}) \\ &\le \frac{3}{\lambda^4}. \end{aligned}$$

Also,

$$\begin{aligned} P(|S_k| > \lambda\sigma\sqrt{n}) &\le \frac{E|S_k|^2/\sigma^2}{\lambda^2 n} \\ &\le \frac{k_\lambda}{\lambda^2 n}. \end{aligned}$$

and hence, we get the above convergence (A) is true.

3.4 Weak convergence on the Skorokhod space

3.4.1 The space $D[0,1]$

Let

$$x : [0,1] \to R$$

be right-continuous with left limit such that

1. for $0 \le t < 1$

$$\lim_{s\searrow t} x(s) = x(t+) = x(t)$$

2. for $0 < t \leq 1$

$$\lim_{s \nearrow t} x(s) = x(t-).$$

We say that $x(t)$ has discontinuity of the first kind at t if left and right limit exist.

For $x \in D$ and $T \subset [0, 1]$,

$$w_x(T) = w(x, T) = \sup_{s,t \in T} |x(s) - x(t)|.$$

We define the modulus of continuity

$$w_x(\delta) = \sup_{0 \leq t \leq 1-\delta} w_x([t, t+\delta))$$

Lemma 3.2 (D1): For each $x \in D$ and $\epsilon > 0$, there exist points $0 = t_0 < t_1 < \cdots < t_\nu = 1$ and $w_x([t_{i-1}, t_i)) < \epsilon$.

Proof: Call t_i's above δ-sparse. If $\min_i\{(t_i - t_{i-1})\} \geq \delta$, define for $0 < \delta < 1$

$$w'_x(\delta) = w'(x, \delta) = \inf_{\{t_i\}} \max_{1 \leq i \leq \nu} w_x([t_{i-1}, t_i)).$$

(If we prove the above lemma, we get $x \in D$, $\lim_{\delta \to 0} w'_x(\delta) = 0$.)

If $\delta < \frac{1}{2}$, we can split $[0, 1)$ into subintervals $[t_{i-1}, t_i)$ such that

$$\delta < (t_i - t_{i-1}) \leq 2\delta$$

and hence,

$$w'_x(\delta) \leq w_x(2\delta).$$

Let us define jump function

$$j(x) = \sup_{0 \leq t \leq 1} |x(t) - x(t-)|.$$

We shall prove that

$$w_x(\delta) \leq 2w'_x(\delta) + j(x).$$

Choose δ-sparse sequence $\{t_i\}$ such that

$$w_x([t_{i-1}, t_i)) < w'_x(\delta) + \epsilon.$$

We can do this from the definition

$$w'_x(\delta) = w'(x,\delta) = \inf_{\{t_i\}} \max_{1\le i\le v} w_x([t_{i-1}, t_i)).$$

If $|s-t| < \delta$, then $s, t \in [t_{i-1}, t_i)$ or belongs to adjoining intervals. Then,

$$|x(s)-x(t)| \begin{cases} w'_x(\delta)+\epsilon, & \text{if } s, t \text{ belong to the same interval;} \\ 2w'_x(\delta)+\epsilon+j(x), & \text{if } s, t \text{ belong to adjoining intervals.} \end{cases}$$

If x is continuous, $j(x) = 0$ and hence,

$$w_x(\delta) \le 2w'_x(\delta).$$

3.4.2 Skorokhod topology

Let Λ be the class of strictly increasing functions on $[0,1]$ and $\lambda(0) = 0$, $\lambda(1) = 1$. Define

$$d(x,y) = \inf\{\epsilon : \exists \lambda \in \Lambda \text{ such that } \sup_t |\lambda(t)-t| < \epsilon \text{ and } \sup_t |x(\lambda(t)) - y(t)| < \epsilon\}.$$

$d(x,y) = 0$ implies there exists $\lambda_n \in \Lambda$ such that $\lambda_n(t) \to t$ uniformly and $x(\lambda_n(t)) \to y(t)$ uniformly. Therefore, with

$$\begin{aligned} ||\lambda - I|| &= \sup_{t\in[0,1]} |\lambda(t)-t| \\ ||x - y\circ\lambda|| &= \sup_{t\in[0,1]} |x(t)-y(\lambda(t))| \\ d(x,y) &= \inf_{\lambda}\left(||\lambda - I|| \vee ||x-y\circ\lambda||\right). \end{aligned}$$

If $\lambda(t) = t$, then
1. $d(x,y) = \sup|x(t)-y(t)| < \infty$ since we showed $|x(s)-x(t)| \le w'_x(\delta) < \infty$.
2. $d(x,y) = d(y,x)$.
3. $d(x,y) = 0$ only if $x(t) = y(t)$ or $x(t) = y(t-)$.

If $\lambda_1, \lambda_2 \in \Lambda$ and $\lambda_1 \circ \lambda_2 \in \Lambda$

$$||\lambda_1\circ\lambda_2 - I|| \le ||\lambda_1 - I|| + ||\lambda_2 - I||.$$

If $\lambda_1, \lambda_2 \in \Lambda$, then the following holds:
1. $\lambda_1 \circ \lambda_2 \in \Lambda$.
2. $||\lambda_1\circ\lambda_2 - I|| \le ||\lambda_1 - I|| + ||\lambda_2 - I||$.

3. $||x - z \circ (\lambda_1 \circ \lambda_2)|| \leq ||x - y \circ \lambda_2|| + ||y - z \circ \lambda_1||$.
4. $d(x,z) \leq d(x,y) + d(y,z)$.

Therefore, Skorokhod topology is given by d.

(D[0,1],d) is not complete. To see this, choose $x_n = 1_{[0,\frac{1}{2^n}]}(t)$ and λ_n be such that $\lambda_n(1/2^n) = 1/2^{n+1}$ and linear in $[0,1/2^n]$ and $[1/2^{n+1},1]$ then $||\, x_{n+1} 0 \lambda_n - x_n \,|| = 0$ and $||\, \lambda_n - I \,|| = \frac{1}{2^{n+1}}$. Meanwhile, $\lambda_n(1/2^n) \neq 1/2^{n+1}$, then $||\, x_n - x_{n+1} \,|| = 1$ and $d(x_n, x_{n+1}) = 1/2^{n+1}$, i.e. x_n is d-Cauchy and $d(x_n, 0) = 1$.

Choose $\lambda \in \Lambda$ near identity. Then for t, s close, $\frac{\lambda(t)-\lambda(s)}{t-s}$ is close to 1. Therefore,

$$||\lambda||^0 = \sup_{s<t} \left| \log \frac{\lambda(t) - \lambda(s)}{t - s} \right| \in (0, \infty).$$

3.5 Metric of $D[0,1]$ to make it complete

Let $\lambda \in \Lambda$ (λ is non-decreasing, $\lambda(0) = 0$, and $\lambda(1) = 1$). Recall

$$||\lambda||^0 = \sup_{s<t} \left| \log \frac{\lambda(t) - \lambda(s)}{t - s} \right| \in (0, \infty).$$

Consider d^0

$$\begin{aligned} d^0(x,y) &= \inf\{\epsilon > 0 : \exists \lambda \in \Lambda \text{ with } ||\lambda||^0 < \epsilon \text{ and } \sup_t |x(t) - y(\lambda(t))| < \epsilon\} \\ &= \inf_{\lambda \in \Lambda}\{||\lambda||^0 \vee ||x - y \circ \lambda||\}. \end{aligned}$$

since, for $u > 0$,

$$|u - 1| \leq e^{|\log u|} - 1,$$

we have

$$\begin{aligned} \sup_{0 \leq t \leq 1} |\lambda(t) - t| &= \sup_{0 \leq t \leq 1} t \left| \frac{\lambda(t) - \lambda(0)}{t - 0} - 1 \right| \\ &= e^{||\lambda||^0} - 1. \end{aligned}$$

For any v, $v \leq e^v - 1$, and hence

$$d(x,y) \leq e^{d^0(x,y)} - 1.$$

Thus, $d^0(x_n, y) \to 0$ implies $d(x_n, y) \to 0$.

Lemma 3.3 (D2): If $x, y \in D[0,1]$ and $d(x,y) < \delta^2$, then $d^0(x,y) \leq 4\delta + w'_x(\delta)$.

Proof: Take $\epsilon < \delta$ and $\{t_i\}$ δ–sparse with

$$w_x([t_{i-1}, t_i)) < w'_x(\delta) + \epsilon \quad \forall i.$$

We can do this from definition of $w'_x(\delta)$. Choose $\mu \in \Lambda$ such that

$$\sup_t |x(t) - y(\mu(t))| = \sup_t |x(\mu^{-1}(t)) - y(t)| < \delta^2 \tag{3.5}$$

and

$$\sup_t |\mu(t) - t| < \delta^2. \tag{3.6}$$

This follows from $d(x, y) < \delta^2$. Take λ to agree with μ at points t_i and linear between. $\mu^{-1} \circ \lambda$ fixes t_i and is increasing in t. Also, $(\mu^{-1} \circ \lambda)(t)$ lies in the same interval $[t_{i-1}, t_i)$. Thus, from (3.5) and (3.6),

$$\begin{aligned} |x(t) - y(\lambda(t))| &\le |x(t) - x((\mu^{-1} \circ \lambda)(t))| + |x((\mu^{-1} \circ \lambda)(t)) - y(\lambda(t))| \\ &= w'_x(\delta) + \epsilon + \delta^2. \end{aligned}$$

$\delta < \frac{1}{2} < 4\delta + w'_x(\delta)$. λ agrees with μ at t_i's. Then by (3.5), (3.6), and $(t_i - t_{i-1}) > \delta$ (δ–sparse),

$$\begin{aligned} |(\lambda(t_i) - \lambda(t_{i-1})) - (t_i - t_{i-1})| &< 2\delta^2 \\ &< 2\delta(t_i - t_{i-1}) \end{aligned}$$

and

$$|(\lambda(t) - \lambda(s)) - (t - s)| \le 2\delta|t - s|$$

for $t, s \in [t_{i-1}, t_i)$ by polygonal property. Now, we take care of adjoining interval. For u_1, u_2, u_3

$$|(\lambda(u_3) - \lambda(u_1)) - (u_3 - u_1)| \le |(\lambda(u_3) - \lambda(u_2)) - (u_3 - u_2)| + |(\lambda(u_2) - \lambda(u_1)) - (u_2 - u_1)|.$$

If t and s are in adjoining intervals, we get the same bound. Since for $u < \frac{1}{2}$

$$|\log(1 \pm u)| \le 2u,$$

we have

$$\log(1 - 2\delta) \le \log \frac{\lambda(t) - \lambda(s)}{t - s} \le \log(1 + 2\delta).$$

Therefore,

$$||\lambda||^0 = \sup_{s<t} \left| \log \frac{\lambda(t) - \lambda(s)}{t - s} \right| < 4\delta$$

and hence, d^0 and d are equivalent. Now, we shall show that D^0 is separable and is complete.

Consider $\sigma = \{s_u\}$ with $0 = s_0 < \cdots < s_k = 1$ and define $A_\sigma : D \to D$ by

$$(A_\sigma x)(t) = x(s_{u-1})$$

for $t \in [s_{u-1}, s_u)$ with $1 \le u \le k$ with $(A_\sigma x)(s_k) = x(1)$.

Lemma 3.4 (D3): If $\max(s_u - s_{u-1}) \le \delta$, then

$$d(A_\sigma x, x) \le \delta \vee w'_x(\delta).$$

Proof: Let $A_\sigma x \equiv \hat{x}$. Let $\zeta(t) = s_{u-1}$ if $t \in [s_{u-1}, s_u)$ with $\zeta(1) = s_k = 1$. Then, $\hat{x}(t) = x(\zeta(t))$. Given $\epsilon > 0$, find δ–sparse set $\{t_i\}$ such that

$$w_x([t_{i-1}, t_i)) < w'_x(\delta) + \epsilon$$

for all i. Let $\lambda(t_i)$ be defined by
1. $\lambda(t_0) = s_0$.
2. $\lambda(t_i) = s_v$ if $t_i \in [s_{v-1}, s_v)$, where

$$t_i - t_{i-1} > \delta \ge s_v - s_{v-1}.$$

Then, $\lambda(t_i)$ is increasing. Now, extend it to $\lambda \in \Lambda$ by linear interpolation.

Claim:

$$\begin{aligned} ||\hat{x}(t) - x(\lambda^{-1}(t)) &= |x(\zeta(t)) - x(\lambda^{-1}(t))| \\ &< w'_x(\delta) + \epsilon. \end{aligned}$$

Clearly, if $t = 0$, or $t = 1$, it is true. Let us look at $0 < t < 1$. First we observe that $\zeta(t), \lambda^{-1}(t)$ lie in the same interval $[t_{i-1}, t_i)$.(We will prove it.) This follows if we shows

$$t_j \le \zeta(t) \text{ iff } t_j \le \lambda^{-1}(t)$$

or equivalently,

$$t_j > \zeta(t) \text{ iff } t_j > \lambda^{-1}(t).$$

This is true for $t_j = 0$. Suppose $t_j \in (s_{v-1}, s_v]$ and $\zeta(t) = s_i$ for some i. By definition, $\zeta(t)$ $t \le \zeta(t)$ is equivalent to $s_v \le t$. Since $t_j \in (s_{v-1}, s_v]$, $\lambda(t_j) = s_v$. This completes the proof.

3.6 Separability of the Skorokhod space

d^0-convergence is stronger than d-convergence.

Theorem 3.5 (D1): The space (D, d) is separable, and hence, so is (D, d^0).

Proof: Let B_k be the set of functions taking constant rational value on $\left[\frac{u-1}{k}, \frac{u}{k}\right]$ and taking rational value at 1. Then, $B = \cup_k B_k$ is countable. Given $x \in D$, $\epsilon > 0$, choose k such that $\frac{1}{k} < \epsilon$ and $w_x\left(\frac{1}{k}\right)$. Apply Lemma D3 with $\sigma = \left\{\frac{u}{k}\right\}$. Note that $A_\sigma x$ has finite many values and

$$d(x, A_\sigma x) < \epsilon.$$

Since $A_\sigma x$ has finitely many real values, we can find $y \in B$ such that given $d(x,y) < \epsilon$,

$$d(A_\sigma x, y) < \epsilon.$$

Now, we shall prove the completeness.

Proof: We take d^0-Cauchy sequence. Then it contains a d^0-convergent subsequence. If $\{x_k\}$ is Cauchy, then there exists $\{y_n\} = \{x_{k_n}\}$ such that

$$d^0(y_n, y_{n+1}) < \frac{1}{2^n}.$$

There exists $\mu_n \in \Lambda$ such that

1. $||\mu_n||^0 < \frac{1}{2^n}$.
2.
$$\begin{aligned}\sup_t |y_n(t) - y_{n+1}(\mu_n(t))| &= \sup_t |y_n(\mu_n^{-1}(t)) - y_{n+1}(t)| \\ &< \frac{1}{2^n}.\end{aligned}$$

We have to find $y \in D$ and $\lambda_n \in \Lambda$ such that

$$||\lambda_n||^0 \to 0$$

and

$$y_n(\lambda_n^{-1}(t)) \to y(t)$$

uniformly.

Heuristic (not a proof): Suppose $y_n(\lambda_n^{-1}(t)) \to y(t)$. Then, by (2), $y_n(\mu_n^{-1}(\lambda_{n+1}^{-1}(t)))$ is within $\frac{1}{2^n}$ of $y_{n+1}(\lambda_{n+1}^{-1}(t))$. Thus, $y_n(\lambda_n^{-1}(t)) \to y(t)$ uniformly.

Find λ_n such that

$$y_n(\mu_n^{-1}(\lambda_{n+1}^{-1}(t))) = y_n(\lambda_n^{-1}(t)),$$

i.e.

$$\mu_n^{-1} \circ \lambda_{n+1}^{-1} = \lambda_n^{-1}.$$

Thus,

$$\begin{aligned}\lambda_n &= \lambda_{n+1}\mu_n \\ &= \lambda_{n+2}\mu_{n+1}\mu_n \\ &\vdots \\ &= \cdots \mu_{n+2}\mu_{n+1}\mu_n.\end{aligned}$$

Proof: Since

$$e^u - 1 \le 2u,$$

for $0 \le u \le \frac{1}{2}$, we have

$$\sup_t |\lambda(t) - t| \le e^{||\lambda||^0} - 1.$$

Therefore,

$$\begin{aligned}\sup_t |(\mu_{n+m+1}\mu_{n+m} \cdots \mu_n)(t) - (\mu_{n+m}\mu_{n+m-1} \cdots \mu_n)(t)| &\le \sup_s |\mu_{n+m+1}(s) - s| \\ &\le 2||\mu_{n+m+1}||^0 \\ &= \frac{1}{2^{n+m}}.\end{aligned}$$

For fixed n,

$$(\mu_{n+m}\mu_{n+m-1} \cdots \mu_n)(t)$$

converges uniformly in t as n goes to ∞. Let

$$\lambda_n(t) = \lim_{m\to\infty} (\mu_{n+m}\mu_{n+m-1} \cdots \mu_n)(t).$$

Then, λ_n is continuous and non-decreasing with $\lambda_n(0) = 0$ and $\lambda_n(1) = 1$. We have to prove $||\lambda_n||^0$ is finite. Then, λ_n is strictly increasing.

$$\left|\log \frac{(\mu_{n+m}\mu_{n+m-1}\cdots\mu_n)(t) - (\mu_{n+m}\mu_{n+m-1}\cdots\mu_n)(s)}{t-s}\right|$$

$$\le ||\mu_{n+m}\mu_{n+m-1}\cdots\mu_n||^0$$

$$(\text{since } \lambda_n \in \Lambda, ||\lambda_n||^0 < \infty)$$

$$\le ||\mu_{n+m}||^0 + \cdots + ||\mu_n||^0$$

$$(\text{since } ||\lambda_1\lambda_2||^0 \le ||\lambda_1||^0 + ||\lambda_2||^0)$$

$$< \frac{1}{2^{n-1}}.$$

Let $m \to \infty$. Then, $||\lambda_n||^0 < \frac{1}{2^{n-1}}$ is finite, and hence, λ_n is strictly increasing. Now, by (2),

$$\begin{aligned}\sup_t |y_n(\lambda_n^{-1}(t)) - y_n(\lambda_{n+1}^{-1}(t))| &\le \sup_s |y_n(s) - y_{n+1}(\mu_n(s))| \\ &< \frac{1}{2^n}.\end{aligned}$$

Therefore, $\{y_n(\lambda_n^{-1}(t))\}$ is Cauchy under supnorm and

$$y_n(\lambda_n^{-1}(t)) \to y(t) \in D$$

and hence converges in d^0.

3.7 Tightness in the Skorokhod space

We turn now to the problem of characterizing compact sets in D. We will prove an analogue of the Arzelà-Ascoli theorem.

Theorem 3.6: A necessary and sufficient condition for a set A to be relatively compact in the Skorohod topology is that

$$\sup_{x\in A} ||x|| < \infty \tag{3.7}$$

and

$$\lim_{\delta\to 0} sup_{x\in A} w'_x(\delta) = 0. \tag{3.8}$$

Proof of sufficiency: Let

$$\alpha = \sup_{x \in A} ||x||.$$

Given $\epsilon > 0$, choose a finite ϵ-net H in $[-\alpha, \alpha]$ and choose δ so that $\delta < \epsilon$ and $w'_x(\delta) < \epsilon$ for all x in A. Apply Lemma 3.4 for any $\sigma = \{s_u\}$ satisfying $\max(s_u - s_{u-1}) < \delta$: $x \in A$ implies $d(x, A_\sigma x) < 2\epsilon$. Take B to be the finite set of y that assume on each $[s_{u-1}, s_u)$ a constant value from H and satisfy $y(1) \in H$. Since B contains a y for which $d(x, A_\sigma x)$, it is a finite 2ϵ-net for A in the sense of d. Thus, A is totally bounded in the sense of d. However, we must show that A is totally bounded in the sense of d^0, since this is the metric under which D is complete. Given (a new) ϵ, choose a new δ so that $0 < \delta \leq 1/2$ and so that $4\delta + w'_x(\delta) < \epsilon$ holds for all x in A. We have already seen that A is d-totally bounded, and thus, there exists a finite set B' that is a δ^2-net for A in the sense of d. However, by Lemma 2, B' is an ϵ-net for A in the sense of d^0.

The proof of necessity requires a Lemma 3.3 and a definition.

Definition 3.3: In any metric space, f is upper semi-continuous at x, if for all $\epsilon > 0$, there exists $\delta > 0$ such that

$$\rho(x, y) < \delta \Rightarrow f(y) < f(x) + \epsilon.$$

Lemma 3.5: For fixed δ, $w'(x, \delta)$ is upper-semicontinuous in x.

Proof: Let x, δ, and ϵ be given. Let $\{t_i\}$ be a δ-spars set such that $w_x[t_{i-1}, t_i) < w'_x(\delta) + \epsilon$ for each i. Now choose η small enough that $\delta + 2\eta < \min(t_i - t_{i-1})$ and $\eta < \epsilon$. Suppose that $d(x, y) < \eta$. Then, for some λ in Λ, we have

$$\sup_t |y(t) - x(\lambda t)| < \eta,$$

$$\sup_t |\lambda^{-1} t - t| < \eta$$

Let $s_i = \lambda^{-1} t_i$. Then $s_i - s_{i-1} > t_i - t_{i-1} - 2\eta > \delta$. Moreover, if s and t both lies in $[s_{i-1}, s_i)$, then λs and λt both lie in $[t_{i-1}, t_i)$, and hence $|y(s) - y(t)| < |x(\lambda s) - x(\lambda t)| + 2\eta \leq w'_x(\delta) + \epsilon + 2\eta$. Thus, $d(x, y) < \eta$ implies $w'_y(\delta) < w'_x(\delta) + 3\epsilon$.

Definition 3.4 (*d*-bounded): A is d-bounded if diameter is bounded, i.e.

$$\text{diameter}(A) = \sup_{x, y \in A} d(x, y) < \infty.$$

Proof of necessity in Theorem 3.6: If A^- is compact, then it is d-bounded, and since $\sup_t |x(t)|$ is the d-distance from x to the 0-function, (3.7) follows. By Lemma 3.1, $w'(x, \delta)$ goes to 0 with δ for each x. However, since $w'(\cdot, \delta)$ is upper-semicontinuous the convergence is uniform on compact sets.

Theorem 3.6, which characterizes compactness in D, gives the following result. Let $\{P_n\}$ be a sequence of probability measure on $(D, \mathcal{D})$.

Theorem 3.7: The sequence $\{P_n\}$ is tight if and only if these two conditions hold: We have

$$\lim_{a\to\infty} \limsup_n P_n\Big(\{x : ||x|| \geq a\}\Big) = 0 \tag{3.9}$$

(ii) for each ϵ,

$$\lim_{\delta\to\infty} \limsup_n P_n\Big(\{x : w'_x(\delta) \geq \epsilon\}\Big) = 0. \tag{3.10}$$

Proof: Conditions (i) and (ii) here are exactly conditions (i) and (ii) of Azela-Ascoli theorem with $||x||$ in place of $|x(0)|$ and w' in place of w. Since D is separable and complete, a single probability measure on D is tight, and so the previous proof goes through.

3.8 The space $D[0, \infty)$

Here we extend the Skorohod theory to the space $D_\infty = D[0, \infty)$ of cadlag functions on $[0, \infty)$, a space more natural than $D = D[0, 1]$ for certain problems.

In addition to D_∞, consider for each $t > 0$ the space $D_t = D[0, t]$ of cadlag functions on $[0, t]$. All the definitions for D_1 have obvious analogues for D_t : $\sup_{s\leq t}|x(s)|$, Λ_t, $||\lambda||_t^0$, d_t^0, d_t. All the theorems carry over from D_1 to D_t in an obvious way. If x is an element of D_∞, or if x is an element of D_u and $t < u$, then x can also be regarded as an element of D_t by restricting its domain of definition. This new cadlag function will be denoted by the same symbol; it will always be clear what domain is intended.

One might try to define the Skorohod convergence $x_n \to x$ in D_∞ by requiring that $d_t^0(x_n, x) \to 0$ for each finite, positive t. However, in a natural theory, $x_n = I_{[0,1-1/n]}$ will converge to $x = I_{[0,1]}$ in D_∞, while $d_1^0(x_n, x) = 1$. The problem here is that x is discontinuous at 1, and the definition must accommodate discontinuities.

Lemma 3.6: Let x_n and x be elements of D_u. If $d_u^0(x_n, x) \to 0$ and $m < u$, and if x is continuous at m, then $d_m^0(x_n, x) \to 0$.

Proof: We can work with the metrics d_u and d_m. By hypothesis, there are elements λ_n of Λ_u such that

$$||\lambda_n - I||_u \to 0$$

and

$$\|x_n - x\lambda_n\|_u \to 0.$$

Given ϵ, choose δ so that $|t - m| \le 2\delta$ implies $|x(t) - x(m)| < \epsilon/2$. Now choose n_0 so that, if $n \ge n_0$ and $t \le u$, then $|\lambda_n t - t| < \delta$ and $|x_n(t) - x(\lambda_n t)| < \epsilon/2$. Then, if $n \ge n_0$ and $|t - m| \le \delta$, we have $|\lambda_n t - m| \le |\lambda_n t - t| + |t - m| < 2\delta$ and hence $|x_n(t) - x(m)| \le |x_n(t) - x(\lambda_n t)| + |x(\lambda_n t) - x(m)| < \epsilon$. Thus

$$\sup_{|t-m|\le\delta} |x(t) - x(m)| < \epsilon, \qquad \sup_{|t-m|\le\delta} |x_n(t) - x(m)| < \epsilon, \qquad \text{for } n \ge n_0. \tag{3.11}$$

If
(i) $\lambda_n m < m$, let $p_n = m - \frac{1}{n}$;
(ii) $\lambda_n m > m$, let $p_n = \lambda^{-1}\left(m - \frac{1}{n}\right)$;
(iii) $\lambda_n m = m$, let $p_n = m$.

Then,
(i) $|p_n - m| = \frac{1}{n}$;
(ii) $|p_n - m| =\le |\lambda_n^{-1}(m - n^{-1}) - (m - n^{-1})| + \frac{1}{n}$;
(iii) $|p_n - m| = m$.

Therefore, $p_n \to m$. Since

$$|\lambda_n p_n - m| \le |\lambda_n p_n - p_n| + |p_n - m|,$$

we also have $\lambda_n p_n \to m$. Define $\mu_n \in \Lambda_n$ so that $\mu_n t = \lambda_n t$ on $[0, p_n]$ and $\mu_n m = m$, and interpolate linearly on $[p_n, m]$. Since $\mu_n m = m$ and μ_n is linear over $[p_n, m]$, we have $|\mu_n t - t| \le |\lambda_n p_n - p_m|$ there, and therefore, $\mu_n t \to t$ uniformly on $[0, m]$. Increase the n_0 of (3.11) so that $p_n > m - \delta$ and $\lambda_n p_n > m - \delta$ for $n \ge n_0$. If $t \le p_n$, then $|x_n(t) - x(\mu_n t)| = |x_n(t) - x(\lambda_n t)| \le \|x_n - x\lambda_n\|_u$. Meanwhile, if $p_n \le t \le m$ and $n \ge n_0$, then $m \ge t \ge p_n > m - \delta$ and $m \ge \mu_n t \ge \mu_n p_n = \lambda_n p_n > m - \delta$, and therefore, by (3.11), $|x_n(t) - x(\mu_n t)| \le |x_n(t) - x(m)| + |x(m) - x(\mu_n t)| < 2\epsilon$. Thus, $|x_n(t) - x(\mu_n t)| \to 0$ uniformly on $[0, m]$.

The metric on D_∞ will be defined in terms of the metrics $d_m^0(x, y)$for integral m, but before restricting x and y to $[0, m]$, we transform them in such a way that they are continuous at m. Define

$$g_n(t) = \begin{cases} 1, & \text{if } t \le m - 1; \\ m - t, & \text{if } m - 1 \le t \le m; \\ 0, & t \ge m. \end{cases} \tag{3.12}$$

For $x \in D_\infty$, let x^m be the element of D_∞ defined by

$$x^m(t) = g_m(t)x(t), \quad t \ge 0 \tag{3.13}$$

Now take

$$d_\infty^0(x,y) = \sum_{m=1}^{\infty} 2^{-m}(1 \wedge d_m^0(x^m, y^m)). \tag{3.14}$$

If $d_\infty^0(x,y) = 0$, then $d_m^0(x,y) = 0$ and $x^m = y^m$ for all m, and this implies $x = y$. The other properties being easy to establish, d_∞^0 is a metric on D_∞; it defines the Skorohod topology there. If we replace d_m^0 by d_m in (3.14), we have a metric d_∞ equivalent to d_∞^0.

Let Λ_∞ be the set of continuous, increasing maps of $[0,\infty)$ onto itself.

Theorem 3.8: There is convergence $d_\infty^0(x_n, x) \to 0$ in D_∞ if and only if there exist elements λ_n of Λ_∞ such that

$$\sup_{t<\infty} |\lambda_n t - t| \to 0 \tag{3.15}$$

and for each m,

$$\sup_{t \le m} |x_n(\lambda_n t) - x(t)| \to 0. \tag{3.16}$$

Proof: Suppose that $d_\infty^0(x_n, x)$ and $d_\infty(x_n, x)$, go to 0. Then there exist elements λ_n^m of Λ_m such that

$$\epsilon_n^m = ||I - \lambda_n^m||_m \vee ||x_n^m \lambda_n^m - x^m||_m \to 0$$

for each m. Choose l_m so that $n \ge l_m$ implies $\epsilon_n^m < 1/m$. Arrange that $l_m < l_{m+1}$, and for $l_m \le n < l_{m+1}$, let $m_n = m$. Since $l_m \le n < l_{m+1}$, we have $m_n \to n$ and $\epsilon_n^{m_n} < 1/m_n$. Define

$$\lambda_n t = \begin{cases} \lambda_n^{m_n} t, & \text{if } t \le m_n; \\ t + \lambda_n^{m_n}(m_n) - m_n, & \text{if } t \ge m_n. \end{cases}$$

Then, $|\lambda_n t - t| < 1/m_n$ for $t \ge m_n$ as well as for $t \le m_n$, and therefore,

$$\sup_t |\lambda_n t - t| \le \frac{1}{m_n} \to 0.$$

Hence, (24). Fix c. If n is large enough, then $c < m_n - 1$, and so

$$||x_n \lambda_n - x||_c = ||x_n^{m_n} \lambda_n^{m_n} - x^{m_n}||_c \le \frac{1}{m_n} \to 0,$$

which is equivalent to (3.16).

Now suppose that (3.15) and (3.16) hold. Fix m. First,

$$x_n^m(\lambda_n t) = g_m(\lambda_n t)x_n(\lambda_n t) \to g_m(t)x(t) = x^m(t) \tag{3.17}$$

holds uniformly on $[0, m]$. Define p_n and μ_n as in the proof of Lemma 1. As before, $\mu_n t \to t$ uniformly on $[0, m]$. For $t \le p_n$, $|x^m(t) - x_n^m(\mu_n t)| = |x^m(t) - x_n^m(\lambda_n t)|$, and this goes to 0 uniformly by (3.17). For the case $p_n \le t \le m$, first note that $|x^m(u)| \le g_m(u)||x||_m$ for all $u \ge 0$ and hence,

$$|x^m(t) - x_n^m(\mu_n t)| \le g_m(t)||x||_m + g_m(\mu_n t)||x_n||_m. \tag{3.18}$$

By (3.15), for large n, we have $\lambda_n(2m) > m$ and hence $||x_n||_m \le ||x_n\lambda_n||_{2m}$; and $||x_n\lambda_n||_{2m} \to ||x_n||_{2m}$ by (3.16). This means that $||x_n||_m$ is bounded(m is fixed). Given ϵ, choose n_0 so that $n \ge n_0$ implies that p_n and $\mu_n p_n$ both lies in $(m-\epsilon, m]$, an interval on which g_m is bounded by ϵ. If $n \ge n_0$ and $p_n \le t \le m$, then t and $\mu_n t$ both lie in $(m - \epsilon, m]$, and it follows by (3.18) that $|x^m(t) - x_n^m(\mu_n t)| \le \epsilon(||x||_m + ||x_n||_m)$. Since $||x_n||_m$ is bounded, this implies that $|x^m(t) - x_n^m(\mu_n t)| \to 0$ holds uniformly on $[p_n, m]$ as well as on $[0, p_n]$. Therefore, $d_m^0(x_n^m, x^m) \to 0$ for each m and hence $d_\infty^0(x_n, x)$ and $d_\infty(x_n, x)$ go to 0. This completes the proof.

Theorem 3.9: There is convergence $d_\infty^0(x_n, x) \to 0$ in D_∞ if and only if $d_t^0(x_n, x) \to 0$ for each continuity point t of x.

Proof: If $d_\infty^0(x_n, x) \to 0$, then $d_\infty^0(x_n^m, x^m) \to 0$ for each m. Given a continuity point t of x, fix an integer m for which $t < m - 1$. By Lemma 1 (with t and m in the roles of m and u) and the fact that y and y^m agree on $[0, t]$, $d_t^0(x_n, x) = d_t^0(x_n^m, x^m) \to 0$.

To prove the reverse implication, choose continuity points t_m of x in such a way that $t_m \uparrow \infty$. The argument now follows the first part of the proof of (3.8). Choose elements λ_n^m of Λ_{t_m} in such a way that

$$\epsilon_n^m = ||\lambda_n^m - I||_{t_m} \vee ||x_n\lambda_n^m - x||_{t_m} \to 0$$

for each m. As before, define integers m_n in such a way that $m_n \to \infty$ and $\epsilon_n^{m_n} < 1/m_n$, and this time define $\lambda_n \in \Lambda_\infty$ by

$$\lambda_n t = \begin{cases} \lambda_n^{m_n} t, & \text{if } t \le t_{m_n}; \\ t, & \text{if } t \ge t_{m_n}. \end{cases}$$

The $|\lambda_n t - t| \le 1/m_n$ for all t, and if $c < t_{m_n}$, then $||x_n\lambda_n - x||_c = ||x_n\lambda_n^{m_n} - x||_c \le 1/m_n \to 0$. This implies that (3.15) and (3.16) hold, which in turn implies that $d_\infty^0(x_n, x) \to 0$. This completes the proof.

3.8.1 Separability and completeness

For $x \in D_\infty$, define $\psi_m x$ as x^m restricted to $[0, m]$. Then, since $d_m^0(\psi_m x_n, \psi_m x) = d_m^0(x_n^m, x^m)$, ψ_m is a continuous map of D_∞ into D_m. In the product space $\Pi = D_1 \times D_2 \times \cdots$, the metric

$$\rho(\alpha, \beta) = \sum_{m=1}^{\infty} 2^{-m}(1 \wedge d_m^0(\alpha_m, \beta_m))$$

defines the product topology, that of coordinatewise convergence. Now define $\psi : D_\infty \to \Pi$ by $\psi x = (\psi_1 x, \psi_2 x, \ldots)$:

$$\psi_m : D_\infty \to D_m, \quad \psi : D_\infty \to \Pi.$$

Then $d_\infty^0(x, y) = \rho(\psi x, \psi y)$: ψ is an isometry of D_∞ into Π.

Lemma 3.7: The image ψD_∞ is closed in Π.

Proof: Suppose that $x_n \in D_\infty$ and $\alpha \in \Pi$ and $\rho(\psi x_n, \alpha) \to 0$; then $d_m^0(x_n^m, \alpha_m) \to 0$ for each m. We must find an x in D_∞ such that $\alpha = \psi x$-that is, $\alpha_m = \psi_m x$ for each m. Let T be the dense set of t such that for every $m \geq t$, α_m is continuous at t. Since $d_m^0(x_n^m, \alpha_m) \to 0$, $t \in T \cap [0, m]$ implies $x_n^m(t) = g_n(t)x_n(t) \to \alpha_m(t)$. This means that for every t in T, the limit $x(t) = \lim x_n(t)$ exists (consider an $m > t + 1$, so that $g_n(t) = 1$). Now $g_m(t)x(t) = \alpha_m(t)$ on $T \cap [0, m]$. It follows that $x(t) = \alpha_m(t)$ on $T \cap [0, m-1]$, so that x can be extended to a cadlag function on each $[0, m - 1]$ and then to a cadlag function on $[0, \infty]$. Now, by right continuity, $g_m(t)x(t) = \alpha_m(t)$ on $[0, m]$, or $\psi_m x = x^m = \alpha_m$. This completes the proof.

Theorem 3.10: The space D_∞ is separable and complete.

Proof: Since Π is separable and complete, so are the closed subspace ψD_∞ and its isometric copy D_∞. This completes the proof.

3.8.2 Compactness

Theorem 3.11: Set A is relatively compact in D_∞ if and only if, for each m, $\psi_m A$ is relatively compact in D_m.

Proof: If A is relatively compact, then $\bar{A}$ is compact and hence the continuous image $\psi_m \bar{A}$ is also compact. But then, $\psi_m A$, as a subset of $\psi_m \bar{A}$, is relatively compact.

Conversely, if each $\psi_m A$ is relatively compact, then each $\overline{\psi_m A}$ is compact, and therefore, $B = \overline{\psi_1 A} \times \overline{\psi_2 A} \times \cdots$ and $E = \psi D_\infty \cap B$ are both compact in Π. But $x \in A$

implies $\psi x \in \overline{\psi_m A}$ for each m, so that $\psi x \in B$. Hence $\psi A \subset E$, which implies that ψA is totally bounded and so is its isometric image A. This completes the proof.

For an explicit analytical characterization of relative compactness, analogous to the Arzela-Ascoli theorem, we need to adapt the $w'(x, \delta)$ to D_∞. For an $x \in D_m$, define

$$w'_m(x, \delta) = \inf \max_{1 \le i \le v} w(x, [t_{i-1}, t_i)), \tag{3.19}$$

where the infimum extends over all decompositions $[t_{i-1}, t_i)$, $1 \le i \le v$, of $[0, m)$ such that $t_t - t_{i-1} > \delta$ for $1 \le i \le v$. Note that the definition does not require $t_v - t_{v-1} > \delta$; Although 1 plays a special role in the theory of D_1, the integers m should play no special role in the theory of D_∞.

The exact analogue of $w'(x, \delta)$ is (3.19), but with the infimum extending only over the decompositions satisfying $t_t - t_{i-1} > \delta$ for $i = v$ as well as for $i < v$. Call this $\bar{w}_m(x, \delta)$. By an obvious extension, a set B in D_m is relatively compact if and only if $\sup_x \|x\|_m < \infty$ and $\lim_\delta \sup_x \bar{w}(x, \delta) = 0$. Suppose that $A \subset D_\infty$ and transform the two conditions by giving $\psi_m A$ the role of B. By (Theorem 3.11), A is relatively compact if and only if, for every m

$$\sup_{x \in A} \|x^m\|_m < \infty \tag{3.20}$$

and

$$\lim_{\delta \to 0} \sup_{x \in A} \bar{w}_m(x^m, \delta) = 0. \tag{3.21}$$

The next step is to show that (3.20) and (3.21) are together equivalent to the condition that, for every m,

$$\sup_{x \in A} \|x\|_m < \infty \tag{3.22}$$

and

$$\lim_{\delta \to 0} \sup_{x \in A} w'_m(x, \delta) = 0. \tag{3.23}$$

The equivalence of (3.20) and (3.22) follows easily because $\|x^m\|_m \le \|x\|_m \le \|x^{m+1}\|_{m+1}$. Suppose (3.22) and (3.23) both hold, and let K_m be the supremum in (3.22). If $x \in A$ and $\delta < 1$, then we have $|x^m(t)| \le K_m \delta$ for $m - \delta \le t < m$. Given ϵ, choose δ so that $K_m \delta < \epsilon/4$ and the supremum in (3.23) is less than $\epsilon/2$. If $x \in A$ and $m - \delta$ lies in the interval $[t_{j-1}, t_j)$ of the corresponding partition, replace the intervals $[t_{i-1}, t_i)$ for $i \ge j$ by the single interval $[t_{j-1}, m)$. This new partition shows that $\bar{w}_m(x, \delta)$. Hence, (3.21).

That (3.21) implies (3.23) is clear because $w'_m(x, \delta) \leq \bar{w}_m(x, \delta)$: An infimum increases if its range is reduced. This gives us the following criterion.

Theorem 3.12: A set $A \in D_\infty$ is relatively compact if and only if (3.22) and (3.23) hold for all m.

3.8.3 Tightness

Theorem 3.13: The sequence $\{P_n\}$ is tight if and only if there two conditions hold:
(i) For each m

$$\lim_{a \to \infty} \limsup_n P_n\Big(\{x : ||x||_m \geq a\}\Big) = 0. \tag{3.24}$$

(ii) For each m and ϵ,

$$\lim_{\delta} \limsup_n P_n\Big(\{x : w'_m(x, \delta) \geq \epsilon\}\Big) = 0. \tag{3.25}$$

There is the corresponding corollary. Let

$$j_m(x) = \sup_{t \leq m} |x(t) - x(t-)|. \tag{3.26}$$

Corollary 3.1: Either of the following two conditions can be substituted for (i) in Theorem 3.13:
(i') For each t in a set T that is dense in $[0, \infty)$,

$$\lim_{a \to \infty} \limsup_n P_n\Big(\{x : |x(t)| \geq a\}\Big) = 0. \tag{3.27}$$

(ii') The relation (3.27) holds for $t = 0$, and for each m,

$$\lim_{a \to \infty} \limsup_n P_n\Big(\{x : j_m(x) \geq a\}\Big) = 0. \tag{3.28}$$

Proof: The proof is almost the same as that for the corollary to Theorem 3.7.

Assume (ii) and (i'). Choose points t_i such that $0 = t_0 < t_1 < \cdots < t_v = m$, $t_i - t_{i-1} > \delta$ for $1 \leq i \leq v - 1$, and $w_x[t_{i-1}, t_i) < w'_m(x, \delta) + 1$ for $1 \leq i \leq v$. Choose from T points s_j such that $0 = s_0 < s_1 < \cdots < s_k = m$ and $s_j - s_{j-1} < \delta$ for $1 \leq j \leq k$. Let $m(x) = \max_{0 \leq j \leq k} |x(s_j)|$. If $t_v - t_{v-1} > \delta$, then $||x||_m \leq m(x) + w'_m(x, \delta) + 1$, just as before. If $t_v - t_{v-1} \leq \delta$(and $\delta < 1$, so that $t_{v-1} > m-1$), then $||x||_{m-1} \leq m(x) + w'_m(x, \delta) + 1$. The old argument now gives (3.24), but with $||x||_m$ replaced by $||x||_{m-1}$, which is just as good.

In the proof that (ii) and (i') imply (i), we have $(v - 1)\delta \leq m$ instead of $v\delta < 1$. However, $v \leq m\delta^{-1} + 1$, and the old argument goes through. This completes the proof.

3.8.4 Aldous's tightness criterion

Consider two conditions, assuming (3.24).

Condition 1^0. For each ϵ, η, m, there exist a δ_0 and an n_0 such that, if $\delta \le \delta_0$ and $n \ge n_0$, and if τ is a discrete X^n–stopping time satisfying $\tau \le m$, then

$$P\Big(|X^n_{\tau+\delta} - X^n_\tau| \ge \epsilon\Big) \le \eta. \tag{3.29}$$

Condition 2^0. For each ϵ, η, m, there exist a δ and an n_0 such that, if $n \ge n_0$, and if τ_1 and τ_2 are a discrete X^n–stopping time satisfying $0 \le \tau_1 \le \tau_2 \le m$, then

$$P\Big(|X^n_{\tau_2} - X^n_{\tau_1}| \ge \epsilon, \tau_2 - \tau_1 \le \delta\Big) \le \eta. \tag{3.30}$$

Theorem 3.14: Conditions 1^0 and 2^0 are equivalent.

Proof: Note that $\tau + \delta$ is a stopping time since

$$\{\tau + \delta \le t\} = \{\tau \le t - \delta\} \in \mathcal{F}^{X_n}_t.$$

In Condition 2^0, put $\tau_2 = \tau, \tau_1 = \tau$. Then it gives Condition 1^0. For the converse, suppose that $\tau \le m$ and choose δ_0 so that $\delta \le 2\delta_0$ and $n \ge n_0$ together imply (3.29). Fix an $n \ge n_0$ and a $\delta \le \delta_0$, and let (enlarge the probability space for X^n) θ be a random variable independent of $\mathcal{F}^n = \sigma(X^s_n : s \ge 0)$ and uniformly distributed over $J = [0, 2\delta]$. For the moment, fix an x in D_∞ and points t_1 and t_2 satisfying $0 \le t_1 \le t_2$. Let μ be the uniform distribution over J, and let $I = [0, \delta]$, $M_i = \{s \in J : |x(t_i + s) - x(t_i)| < \epsilon\}$, and $d = t_2 - t_1$.

Suppose that

$$t_2 - t_1 \le \delta \tag{3.31}$$

and

$$\mu(M_i) = P(\theta \in M_i) > \frac{3}{4}, \quad \text{for } i = 1, 2 \tag{3.32}$$

If $\mu(M_2 \cap I) \le \frac{1}{4}$, then $\mu(M_2) \le \frac{3}{4}$, which is a contradiction. Hence, $\mu(M_2 \cap I) > \frac{1}{4}$, and for $d (0 \le d \le \delta)$, $\mu((M_2 + d) \cap J) \le \mu((M_2 \cap I) + d) = \mu((M_2 \cap I))\frac{1}{4}$. Thus $\mu(M_1) + \mu((M_2 + d) \cap J) > 1$, which implies $\mu(M_1 \cap (M_2 + d)) > 0$. There is therefore an s such that $s \in M_1$ and $s - d \in M_2$, from which follows

$$|x(t_1) - x(t_2)| < 2\epsilon. \tag{3.33}$$

Thus, (3.31) and (3.32) together implies (3.33). To put it another way, if (3.31) holds but (3.33) does not, then either $P(\theta \in M_1^c) \geq \frac{1}{4}$ or $P(\theta \in M_2^c) \geq \frac{1}{4}$. Therefore,

$$P\Big(|X^n_{\tau_2} - X^n_{\tau_1}| \geq 2\epsilon, \tau_2 - \tau_1 \leq \delta\Big) \leq \sum_{i=1}^{2} P\Big[P\Big(|X^n_{\tau_i+\theta} - X^n_{\tau_i}| \geq \epsilon | \mathcal{F}^n\Big) \geq \frac{1}{4}\Big]$$
$$\leq 4 \sum_{i=1}^{2} P\Big(|X^n_{\tau_i+\theta} - X^n_{\tau_i}| \geq \epsilon\Big).$$

Since $0 \leq \theta \leq 2\delta \leq 2\delta_0$, and since θ and $\mathcal{F}^n$ are independent, it follows by (3.29) that the final term here is at most 8η. Therefore, Condition 1^0 implies Condition 2^0.

This is Aldous's theorem:

Theorem 3.15 (Aldous): If (3.24) and Condition 1° hold, then $\{X^n\}$ is tight.

Proof: By Theorem 3.13, it is enough to prove that

$$\lim_{a \to \infty} \limsup_{n} P\Big(w'_m(X^n, \delta) \geq \epsilon\Big) = 0. \tag{3.34}$$

Let Δ_k be the set of nonnegative dyadic rationals[1] $j/2^k$ of order k. Define random variables $\tau_0^n, \tau_1^n, \ldots$ by $\tau_0^n = 0$ and

$$\tau_i^n = \min\{t \in \Delta_k : \tau_{i-1}^n < t \leq m, |X_t^n - X^n_{\tau^n_{i-1}}| \geq \epsilon\},$$

with $\tau_i^n = m$ if there is no such t. The τ_i^n depend on ϵ, m, and k as well as on i and n, although the notation does not show this. It is easy to prove by induction that the τ_i^n are all stopping times.

Because of Theorem 3.14, we can assume that condition 2 holds. For given ϵ, η, m, choose δ' and n_0 so that

$$P\Big(|X^n_{\tau_i} - X^n_{\tau_{i-1}}| \geq \epsilon, \tau_i^n - \tau_{i-1}^n \leq \delta'\Big) \leq \eta$$

for $i \geq 1$ and $n \geq n_0$. Since $\tau_i^n < m$ implies that $|X^n_{\tau_i} - X^n_{\tau_{i-1}}| \geq \epsilon$, we have

$$P\Big(\tau_i^n < m, \tau_i^n - \tau_{i-1}^n \leq \delta'\Big) \leq \eta, \quad i \geq 1, n \geq n_0. \tag{3.35}$$

Now choose an integer q such that $q\delta \geq 2m$. There is also a δ such that

$$P\Big(\tau_i^n < m, \tau_i^n - \tau_{i-1}^n \leq \delta\Big) \leq \frac{\eta}{q}, \quad i \geq 1, n \geq n_0. \tag{3.36}$$

1 Dyadic rational is a rational number whose denominator is a power of 2, i.e. a number of the form $a/2b$, where a is an integer and b is a natural number; for example, 1/2 or 3/8, but not 1/3. These are precisely the numbers whose binary expansion is finite.

However,

$$P\Big(\bigcup_{i-1}^{q}\{\tau_i^n < m, \tau_i^n - \tau_{i-1}^n \le \delta\}\Big) \le \eta, \quad n \ge n_0. \tag{3.37}$$

Although τ_i^n depends on k, (3.35) and (3.37) hold for all k simultaneously. By (3.35),

$$\begin{aligned} E(\tau_i^n - \tau_{i-1}^n|\tau_q^n < m) &\ge \delta' P(\tau_i^n - \tau_{i-1}^n \ge \delta'|\tau_q^n < m) \\ &\ge \delta'\Big(1 - \frac{\eta}{P(\tau_q^n < m)}\Big), \end{aligned}$$

and therefore,

$$\begin{aligned} m &\ge E(\tau_q^n|\tau_q^n < m) \\ &= \sum_{i=1}^{q} E(\tau_i^n - \tau_{i-1}^n|\tau_q^n < m) \\ &\ge q\delta'\Big(1 - \frac{\eta}{P(\tau_q^n < m)}\Big). \end{aligned}$$

Since $q\delta' \ge 2m$ by the choice of q, this leads to $P(\tau_q^n < m) \ge 2\eta$. By this and (3.37),

$$P\Big(\{\tau_q^n < m\} \cup \bigcup_{i-1}^{q}\{\tau_i^n < m, \tau_i^n - \tau_{i-1}^n \le \delta\}\Big) \le 3\eta, \quad k \ge 1, n \ge n_0 \tag{3.38}$$

Let A_{n_k} be the complement of the set in (3.38). On this set, let v be the first index for which $\tau_v^n = m$. Fix an n beyond n_0. There are points $t_i^k(\tau_i^n)$ such that $0 = t_0^k < \cdots < t_v^k = m$ and $t_i^k - t_{i-1}^k > \delta$ for $1 \le i < v$. $|X_t^n - X_s^n| < \epsilon$ if s, t lie in the same $[t_{i-1}^k, t_i^k)$ as well as in Δ_k. If $A_n = \limsup_k A_{n_k}$, then $P(A_n) \ge 1 - 3\eta$, and on A_n there is a sequence of values of k along which v is constant ($v \le q$), and for each $i \le v$, t_i^k converges to some t_i. However, $0 = t_0 < \cdots < t_v = m$, $t_i - t_{i-1} \ge \delta$ for $i < v$, and by right continuity, $|X_t^n - X_s^n| \le \epsilon$ if s, t lie in the same $[t_{i-1}, t_i)$. It follows that $w'(X^n, \delta) \le \epsilon$ on a set of probability at least $1 - 3\eta$ and hence (3.34).

As a corollary to the theorem we get.

Corollary 3.2: If for each m, the sequences $\{X_0^n\}$ and $\{j_m(X^n)\}$ are tight on the line, and if Condition 1° holds, then $\{X^n\}$ is tight.

Skorokhod introduced the topology on $D[0,\infty)$ to study the weak convergence of processes with independent increment. Semi-martingales are generalizations of these processes. In the next chapter, we shall discuss the weak convergence of semi-martingales.

4 Central limit theorem for semi-martingales and applications

As stated at the end of chapter 3, Skorokhod developed the theory of weak convergence of stochastic processes with values in $D[0, T]$ to consider the limit theorems (or invariance principle) with convergence to process with independent increments. Since these are special classes of semi-martingales that have sample paths in $D[0, \infty)$, we now study the work of Liptser and Shiryaev [17] (see [15]) on weak convergence of a sequence of semi-martingales to a limit. We begin with the definition of semi-martingales and their structure, including semi-martingale characteristics [15]. Based on this, we obtain conditions for the weak convergence of semi-martingale sequence to a limit. We begin with some preliminary lemmas, which will be needed in the proofs. We end the chapter by giving applications to statistics of censored data that arises in survival analysis in clinical trials.

4.1 Local characteristics of semi-martingale

In this chapter, we study the central limit theorem by Lipster and Shiryayev. We begin by giving some preliminaries. We consider $(\Omega, \{\mathcal{F}_t\}, \mathcal{F}, P)$ a filtered probability space, where $\mathbf{F} = \{\mathcal{F}_t, t \geq 0\}$ is a non-decreasing family of sub σ-field of $\mathcal{F}$, satisfying $\bigcap_{t \geq s} \mathcal{F}_t = \mathcal{F}_s$. We say that $\{\underline{X}, \mathbf{F}\}$ is a martingale if for each t, $X_t \in \underline{X} = \{X_t\} \subset L_1(\Omega, \mathcal{F}_t, P)$ and $E(X(t)|\mathcal{F}_s) = X_s$ a.e. P. WLOG, we assume $\{X_t, t \geq 0\}$ is $D[0, \infty)$ valued (or a.s. it is cadlag) as we can always find a version. A martingale X is said to be square-integrable if $\sup_t EX_t^2 < \infty$. We say that $\{X_t, t \geq 0\}$ is locally square integrable martingale if there exists an increasing sequence σ_n of $(\mathcal{F}_t)$-stopping times such that $0 \leq \sigma_n < \infty$ a.e. $\lim_n \sigma_n = \infty$, and $\{X(t \wedge \sigma_n)1_{\{\sigma_n > 0\}}\}$ is a square integrable martingale. A process $(\underline{X}, \mathbf{F})$ is called a semi-martingale if it has the decomposition

$$X_t = X_0 + M_t + A_t,$$

where $\{M_t\}$ is local martingale, $M_0 = 0$, A is right continuous process with A_0, A_t, $\mathcal{F}_t$-measurable and has sample paths of finite variation. We now state condition of A to make this decomposition unique (called canonical decomposition). For this we need the following. We say that a sub σ-field of $[0, \infty) \times \Omega$ generated by sets of the form $\{(s, t] \times A, 0 < s \leq t < \infty$ and $A \in \mathcal{F}_s\}$ and $\{0\} \times B (B \in \mathcal{F}_0)$ is the σ-field of predictable sets from now on called predictable σ-algebra $\mathcal{P}$. In the above decomposition, A is $\mathcal{P}$-measurable then it is canonical.

We remark that if the jumps of the semi-martingale are bounded, then the decomposition is canonical.

We now introduce the concept of local characteristics.

Let $(\underline{X}, \mathbf{F})$ be a semi-martingale. Set with $\Delta X(s)$ as jump at s,

$$\tilde{X}(t) = \sum_{s \le t} \Delta X(s) 1(|\Delta X(s)| \ge \epsilon).$$

The $\underline{X}(t) = X(t) - \tilde{X}(t)$ is a semi-martingale with unique canonical decomposition

$$\underline{X}(t) = X(0) + M(t) + A(t),$$

where $(M, \mathbf{F})$ is a local martingale and A is predictable process of finite variation. Thus,

$$X(t) = X(0) + M(t) + A(t) + \tilde{X}(t).$$

Let

$$\mu((0, t]; A \cap \{|x| \ge \epsilon\}) = \sum 1(\Delta X(s) \in A, |\Delta X(s)| > \epsilon)$$

and $\nu((0, t]; \cdot \cap \{|x| \ge \epsilon\})$ its predictable projection. Then for each $\epsilon > 0$, we can write with ν as a mesaure generated on $\nu(*x0)$

$$X(t) = X(0) + M'(t) + A(t) + \int_0^1 \int_{|x|>\epsilon} x\nu(ds, dx),$$

where $(M', \mathbf{F})$ is a local martingale. Now the last two terms are predictable. Thus, the semi-martingale is described by M', A, and ν. We thus have the following.

Definition 4.1: The local characteristic of a semi-martingale X is defined by the triplet (A, C, ν), where

1. A is the predictable process of finite variation appearing in the above decomposition of $\underline{X}(t)$.
2. C is a continuous process defined by C_t

$$C_t = [\underline{X}, \underline{X}]_t^c = \langle M^c \rangle_t.$$

3. ν is the predictable random measure on $R_+ \times R$, the dual predictable projection of the measure μ associated to the jumps of X given on $(\{0\}^c)$ by

$$\mu(w, dt, dx) = \sum_{s>0} 1(\Delta X(s, w) \ne 0)\delta_{(s, \Delta X(s,w))}(dt, dx)$$

with δ being the dirac delta measure.

4.2 Lenglart inequality

Lemma 4.1 Mcleish lemma: Let $F_n(t), n = 1, 2, \ldots$ and $F(t)$ be in $D[0, \infty)$ such that $F_n(t)$ is increasing in t for each n, and $F(t)$ is a.s. continuous and for each $t > 0$, there exists $t_n \to t$ such that $F_n(t_n) \to F(t)$. Then,

$$\sup_t |F_n(t) - F(t)| \to 0,$$

where sup is taken over a compact set of $[0, \infty)$.

Proof: WLOG, choose compact set $[0, 1]$: For $\epsilon > 0$ choose $\{t_{n_i}, i = 0, 1, \ldots, k\}$ for fixed $k \geq 1/\epsilon$ such that $t_{n_i} \to i\epsilon$ and $F_n(t_{n_i}) \to_p F(i\epsilon)$ as $n \to \infty$. Then

$$\begin{aligned} \sup_t |F_n(t) - F(t)| \leq \sup_i |F_n(t_{n_{i+1}}) - F_n(t_{n_i})| + \sup_i |F_n(t_{n_i}) - F(t_{n_i})| \\ + \sup_i |F_n(t_{n_{i+1}}) - F(t_{n_i})| + \epsilon. \end{aligned}$$

As $n \to \infty$, choose ϵ such that $|F((i+1)\epsilon) - F(i\epsilon)|$ is small.

We assume that A_t is an increasing process for each t and is $\mathcal{F}_t$-measurable.

Definition 4.2: An adapted positive right continuous process X_t is said to be dominated by an increasing predictable process A if for all finite stopping times T we have

$$EX_T \leq EA_T$$

Example: Let M_t^2 is square martingale. Consider $X_t = M_t^2$. Then, we know $X_t - < M >_t$ is a martingale, and hence, $X_T - < M >_T$ is a martingale. Thus,

$$\begin{aligned} E(X_T - < M >_T) = EX_0 = 0 \\ \Rightarrow EX_T = E < M >_T \end{aligned}$$

Let

$$X_t^* = \sup_{s \leq t} |X_s|$$

Lemma 4.2: Let T be a stopping time and X be dominated by increasing process A (as above). Then,

$$P(X_T^* \geq c) \leq \frac{E(A_T)}{c}$$

for any positive c.

Proof: Let $S = \inf\{s \le T \wedge n : X_s \ge c\}$, clearly $S \le T \wedge n$. Thus,

$$\begin{aligned} EA_T &\ge EA_S \text{ (since } A \text{ is an increasing process)} \\ &\ge EX_S \text{ (since } X \text{ is dominated by } A) \\ &\ge \int_{\{X^*_{T\wedge n} > c\}} X_S dP \text{ (since } X_S > 0 \text{ on } \{X^*_{T\wedge n} > c\}) \\ &\ge c \cdot P(X^*_{T\wedge n} > c) \end{aligned}$$

Therefore, we let n go to ∞, then we get

$$EA_T \ge c \cdot P(X^*_T > c).$$

Theorem 4.1 (Lenglart Inequality): If X is dominated by a predictable increasing process, then for every positive c and d

$$P(X^*_T > c) \le \frac{E(A_T \wedge d)}{c} + P(A_T > d).$$

Proof: It is enough to prove for predictable stopping time $T > 0$,

$$P(X^*_{T-} \ge c) \le \frac{1}{c} E(A_{T-} \wedge d) + P(A_{T-} \ge d). \tag{4.1}$$

We choose $T' = \infty$. Then T' is predictable, $\sigma_n = n$ and apply to $X^T_t = X_{t\wedge T}$ for T finite stopping time $X^{T^*}_{T-} = X^*_T$.

To prove (4.1)

$$\begin{aligned} P(X^*_{T-} \ge c) &= P(X^*_{T-} \ge c, A_{T-} < d) + P(X^*_{T-} \ge c, A_{T-} \ge d) \\ &\le P(X^*_{T-} \ge c) + P(A_{T-} \ge d) \end{aligned} \tag{4.2}$$

Let $S = \inf\{t : A_t \ge d\}$. It is easy to show that S is a stopping time. Also, S is predictable.

On $\{\omega : A_{T-} < d\}$, $S(\omega) \ge T(\omega)$, and hence

$$1(A_{t-} < d) X^*_{T-} \le X^*_{(T\wedge S)-}.$$

By (4.2), we have

$$\begin{aligned} P(X^*_{T-} \ge c) &\le P(X^*_{T-} \ge c) + P(A_{T-} \ge d) \\ &\le P(X^*_{T\wedge S} \ge c) + P(A_{T-} \ge d) \end{aligned}$$

Let $\epsilon > 0$, $\epsilon < c$ and $S_n \nearrow S \wedge T$. Then,

$$\begin{aligned} P(X^*_{(T\wedge S)-} \geq c) &\leq \liminf_n P(X^*_{S_n} \geq c - \epsilon) \text{ (by Fatou's lemma)} \\ &\leq \frac{1}{c-\epsilon} \lim_{n\to\infty} EA_{S_n} \text{ (by Lemma 4.2)} \\ &= \frac{1}{c-\epsilon} EA_{(S\wedge T)} \text{ (by Monotone convergence theorem)} \end{aligned}$$

Since ϵ is arbitrary,

$$\begin{aligned} P(X^*_{(T\wedge S)-} \geq c) &\leq \frac{1}{c} EA_{S_n} \leq \frac{1}{c} E(A_{(S\wedge T)-}) \\ &\leq \frac{1}{c} E(A_{(T-\wedge d)}) \end{aligned}$$

This completes the proof.

Corollary 4.1: Let $M \in \mathcal{M}^2_{LOC}((\mathcal{F}_t), P)$ (class of locally square integrable martingale). Then,

$$P(\sup_{t\leq T} |M_t| > a) \leq \frac{1}{a^2} E(< M >_T \wedge b) + P(< M >_T \geq b)$$

Proof: Use $X_t = |M_t|^2$, $c = a^2$, $b = d$, $A_t = < M >_t$.

Lemma 4.3: Consider $\{\mathcal{F}^n_t, P\}$. Let $\{M^n\}$ be locally square martingale. Assume that

$$< M^n >_t \longrightarrow_p f(t),$$

where f is continuous and deterministic function (hence, f will be an increasing function). Then, $\{P \circ (M^n)^{-1}\}$ on $D[0, \infty)$ is relatively compact on $D[0, \infty)$.

Proof: It suffices to show for constant $T < \infty$, and for any $\eta > 0$ there exists $a > 0$ such that

$$\sup_n P(\sup_{t\leq T} |M^n_t| > a) < \eta. \tag{4.3}$$

For each $T < \infty$ and $\eta, \epsilon > 0$, there exist n_0, δ such that for any stopping time τ^n(w.r.t $(\mathcal{F}^n_t)$, $\tau^n \leq T, \tau^n + \delta < T$).

$$\sup_{n\geq n_0} P(\sup_{0\leq t\leq\delta} |M^n_{\tau^n+t} - M^n_{\tau_n}| \geq \epsilon) < \eta. \tag{4.4}$$

Observe that by corollary to Lenglart inequality,

$$P(\sup_{t\leq T}|M_t^n| > a) \leq \frac{1}{a^2}E(< M^n >_T \wedge b) + P(< M^n >_T \geq b)$$

Let $b = f(T) + 1$, then under the hypothesis, there exists n_1 such that for all $n \geq n_1$

$$P(< M^n >_T \geq b) < \frac{\eta}{2}.$$

Thus, for $n \geq n_1$

$$\sup_n P(\sup_{t\leq T}|M_t^n| > a) \leq \frac{b}{a^2} + \frac{\eta}{2} + \sum_{k=1}^{n_1} P(\sup_{t\leq T}|M_t^k| > a).$$

Choose a large to obtain (4.3).

We again note that $M^n_{\tau+t} - M^n_\tau$ is a locally square integrable martingale. Hence, by Corollary 4.1 and triangle inequality

$$\begin{aligned}
P(\sup_{0\leq t\leq\delta}|M^n_{\tau+t} - M^n_t| \geq \epsilon) &\leq \frac{1}{\epsilon^2}E\Big(\Big(< M^n >_{\tau+\delta} - < M^n >_\tau \Big)\wedge b\Big) \\
&\quad + P\Big(< M^n >_{\tau+\delta} - < M^n >_\tau \geq b\Big) \\
&\leq \frac{1}{\epsilon^2}E\Big(\sup_{t\leq T}\Big(< M^n >_{t+\delta} - < M^n >_t \Big)\wedge b\Big) \\
&\quad + P\Big(\sup_{t\leq T}| < M^n >_{t+\delta} - < M^n >_t | \geq b\Big) \\
&\leq \frac{1}{\epsilon^2}E\Big(\sup_{t\leq T}\Big|M^n_{t+\delta} - f(t+\delta)\Big| \wedge b\Big) \\
&\quad + \frac{1}{\epsilon^2}E\Big(\sup_{t\leq T}\Big|M^n_t - f(t+\delta)\Big| \wedge b\Big) \\
&\quad + \frac{1}{\epsilon^2}\sup_{t\leq T}|f(t+\delta) - f(t)| \\
&\quad + P\Big(\sup_{t\leq T}| < M^n >_{t+\delta} - f(t+\delta)| \geq \frac{b}{3} \geq b\Big) \\
&\quad + P\Big(\sup_{t\leq T}| < M^n >_t - f(t)| \geq \frac{b}{3} \geq b\Big) \\
&\quad + 1\Big(\sup_{t\leq T}|f(t+\delta) - f(t)| \geq \frac{b}{3}\Big) \\
&\quad + P\Big(\sup_{t\leq T}| < M^n >_{t+\delta} - < M^n >_t | \geq b\Big).
\end{aligned}$$

Using the McLeish lemma, each term goes to 0. This completes the proof.

If our conditions guarantee that for a locally square integrable martingale the associated increasing process converges, then the problem reduces to the convergence of finite-dimensional distributions for weak convergence of $\{M_t^n\}$ to the convergence of finite-dimensional distributions.

4.3 Central limit theorem for semi-martingale

Theorem 4.2: Let $\{X^n\}$ be a sequence of semi-martingale with characteristics $(B^n[X^n, X^n], \nu^n)$ and M be a continuous Gaussian martingale with increasing process $< M >$.

(i) For any $t > 0$ and $\epsilon \in (0, 1)$, let the following conditions be satisfied:

(A)

$$\int_0^t \int_{|x|>\epsilon} \nu^n(ds, dx) \to_p 0$$

(B)

$$B_t^{nc} + \sum_{0 \le s \le t} \int_{|x| \le \epsilon} x \nu^n(\{s\}, dx) \to_p 0$$

(C)

$$< X^{nc} >_t + \int_0^t \int_{|x| \le \epsilon} x^2 \nu^n(ds, dx) - \sum_{0 \le s \le t} \left(\int_{|x| \le \epsilon} x \nu^n(\{s\}, dx) \right)^2 \to_P < M >_t$$

Then $X^n \Rightarrow M$ for finite dimension.

(ii) If (A) and (C) are satisfied as well as the condition

$$\sup_{0 < s \le t} \left| B_s^{nc} + \sum_{0 \le u \le s} \int_{|x| \le \epsilon} x \nu^n(\{u\}, dx) \right| \to_P 0 \tag{4.5}$$

for any t and $\epsilon \in (0, 1]$, then $X \Rightarrow M$ in $D[0, T]$.

Proof: Let μ^n be as defined in Section 4.1 and υ^n be associated predictable projection. For $\epsilon \in (0, 1]$,

$$\begin{aligned} X_t^n &= \left(\sum_{0 \le s \le t} \int_{\epsilon < |x| \le 1} x \nu^n(\{s\}, dx) \right) + \left(B_t^{nc} + \sum_{0 \le s \le t} \int_{|x| \le \epsilon} x \nu^n(\{s\}, dx) \right) \\ &+ \left(\int_0^t \int_{|x|>1} x \mu^n(ds, dx) + \int_0^t \int_{\epsilon < |x| \le 1} (\mu^n - \nu^n)(ds, dx) \right) \\ &+ \left(X_t^{nc} + \int_0^t \int_{|x| \le o} (\mu^n - \nu^n)(ds, dx) \right) \\ &= \alpha_t^n(\epsilon) + \beta_t^n(\epsilon) + \gamma_t^n(\epsilon) + \Delta_t^n(\epsilon). \end{aligned}$$

where

$$\alpha_t^n(\epsilon) = \sum_{0\le s\le t} \int_{\epsilon<|x|\le 1} x\nu^n(\{s\}, dx)$$

$$\beta_t^n(\epsilon) = B_t^{nc} + \sum_{0\le s\le t} \int_{|x|\le\epsilon} x\nu^n(\{s\}, dx)$$

$$\gamma_t^n(\epsilon) = \int_0^t \int_{|x|>1} x\mu^n(ds, dx) + \int_0^t \int_{\epsilon<|x|\le 1} (\mu^n - \nu^n)(ds, dx)$$

$$\Delta_t^n(\epsilon) = X_t^{nc} + \int_0^t \int_{|x|\le\epsilon} (\mu^n - \nu^n)(ds, dx).$$

By (A) we have

$$\sup_{s\le t} \alpha_s^n(\epsilon) \to_p 0.$$

By (B) we have

$$\beta_t^n(\epsilon) \to_p 0.$$

By (4.3) we have

$$\sup_{s\le t} \beta_s^n(\epsilon) \to_p 0.$$

Let

$$Y_t^n = \gamma_t^n(\epsilon) + \Delta_t^n(\epsilon).$$

It suffices to prove $Y^n \to M$ on $D[0, T]$ for each T(by the decomposition Y^n does not depend on ϵ). Next, we have

$$\sup_{0<t\le T} |\gamma_t^n(\epsilon)| \le \int_0^T \int_{|x|>1} |x|\mu^n(ds, dx) + \int_0^T \int_{|x|>\epsilon} \mu^n(ds, dx) + \int_0^T \int_{|x|>\epsilon} \nu^n(ds, dx). \quad (4.6)$$

Note that the second term on RHS $\to 0$ by Lenglart inequality and the third by (A).

Therefore, if we can show that the first term of RHS goes to 0, then $\sup_{0<t\le T} |\gamma_t^n(\epsilon)| \to 0$. We have

$$\int_0^T \int_{|x|>1} |x|\mu^n(ds, dx) = \sum_{0<s\le T} |\Delta X_s^n| 1_{(|\Delta X_s^n|>1)}.$$

For $\delta \in (0, 1)$,

$$\Big\{ \sum_{0<s\le T} |\Delta X_s^n| 1_{(|\Delta X_s^n|>1)} > \delta \Big\} \subset \Big\{ \sum_{0<s\le T} 1_{(|\Delta X_s^n|>1)} > \delta \Big\}.$$

and

$$\sum_{0<s\leq T} 1_{(|\Delta X_s^n|>1)} = \int_0^T \int_{|x|>1} \mu^n(ds, dx) \to_p 0.$$

by Lenglart inequality. Therefore, by (4.6), we have

$$\sup_{0<t\leq T} |\gamma_t^n(\epsilon)| \to 0.$$

Now, only thing left is to show that

$$\Delta_t^n(\epsilon) \to 0$$

$$\Delta_t^n(\epsilon) = X_t^{nc} + \int_0^T \int_{|x|\leq\epsilon} x(\mu^n - \nu^n)(ds, dx).$$

Since $(\mu^n - \nu^n)$ is martingale and X^{nc} is martingale, Δ^n is martingale. Since $\Delta^n(\epsilon) \in \mathcal{M}_{LOC}((\mathcal{F}^n)_t, P)$,

$$\begin{aligned} < \Delta^n(\epsilon) >_t &= < X^{nc} >_t + \int_0^t \int_{|x|\leq\epsilon} x^n \nu^n(ds, dx) \\ &\quad - \sum_{0<s\leq t} \left(\int_{|x|\leq\epsilon} x\nu^n(\{s\}, dx) \right)^2 \\ &\longrightarrow < M >_t \end{aligned}$$

by condition (C).

By McLeish lemma,

$$\sup_{t\leq T} | < \Delta^n(\epsilon) >_t - < M >_t | \to_p 0. \tag{4.7}$$

We showed $\sup_{t\leq T} |\gamma_t^n(\epsilon)| \to 0$. Combining this with (4.7), we have

$$\max\left(\sup_{t\leq T} | < \Delta^n(\epsilon) >_t - < M >_t |, \sup_{t\leq T} |\gamma_t^n(\epsilon)| \right) \to 0$$

Then, there exists $\{\epsilon_n\}$ such that

$$\sup_{t\leq T} | < \Delta^n(\epsilon_n) >_t - < M >_t | \to 0, \quad \sup_{t\leq T} |\gamma_t^n(\epsilon_n)| \to 0 \tag{4.8}$$

$$M_t^n = \Delta_t^n(\epsilon_n), \quad Y_t^n = \Delta_t^n(\epsilon_n) + \gamma_t^n(\epsilon_n).$$

It suffices to prove that $M^n \Rightarrow M$. $\{M_t^n = \Delta_t^n(\epsilon_n)\}$ is compact by Lemma 4.3 and (4.7). It suffices to prove finite-dimensional convergence.

Let $H(t), 0 \leq t \leq T$ be a piecewise constant left-continuous function assuming finitely many values. Let

$$N_t^n = \int_0^t H(s)dM_s^n, \quad N_t = \int_0^t H(s)dM_s.$$

Since M is Gaussian, N is also Gaussian.

Remark: Cramer-Wold criterion for finite-dimensional convergence

$$Ee^{iN_T^n} \to Ee^{iN_T} = e^{-\frac{1}{2}\int_0^T H^2(s)d<M>_s}.$$

Let A be predictable, $A \in \mathcal{A}_{LOC}(\mathcal{F}, P)$

$$e(A)_t = e^{A_t} \prod_{0 \leq s \leq t} (1 + \Delta A_s)e^{-A_s}.$$

Then e^{A_t} will be a solution of $dZ_t = Z_{t-}dA_t$ by Dolean-Dade's work. If $m \in \mathcal{M}_{LOC}$, then

$$A_t = -\frac{1}{2} < m^c >_t + \int_0^t \int_{R-\{0\}} (e^{isx} - 1 - ix)\nu_m(ds, dx).$$

Lemma 4.4: For some $a > 0$ and $c \in (0, 1)$, let $< m >_\infty \leq a$, $\sup_t |\Delta m_t| \leq c$. Then $(e(A_t), \mathcal{F}_t)$ is such that

$$|e(A)_t| \geq \exp\left(-\frac{2a}{1-c^2}\right),$$

and the process $(Z_t, \mathcal{F}_t)$ with $Z_t = e^{imt}\left(e(A)_t\right)^{-1}$ is a uniformly integrable martingale. We will use the lemma to prove

$$Ee^{iN_T^n} \to Ee^{iN_T} = e^{-\frac{1}{2}\int_0^T H^2(s)d<M>_s}.$$

Case 1: Let us first assume $< M^n >_T \leq a$ and $a > 1+ < M >_T$. Observe that
1. By (4.7), $< N^n >_t \to < N >_t$.
2. $|\Delta_t^n| \leq 2\epsilon_n$.
3. $|\Delta N_t^n| \leq 2\lambda\epsilon_n = d_n$, where $\lambda = \max_{t \leq T} |H(s)|$.

We want to prove

$$E \exp\left(iN_T^n + \frac{1}{2} < N >_T\right) \to 1. \tag{4.9}$$

Let A_t^n be increasing process associated with N_t^n. Let $Z_t = e^{iN_t^n}\left(e(A^n)_t\right)^{-1}$. Choose n_0 such that $d_{n_0} = 2\lambda\epsilon_{n_0} \leq 1/2$. By Lemma 4.4, Z^n is a martingale with $EZ_T^n = 1$. To prove (57) is equivalent to proving

$$\lim_{n\to\infty}\left(E\exp\left(iN_T^n + \frac{1}{2}<N>_T\right) - Ee^{iN_T^n}\left(e(A^n)_T\right)^{-1}\right) = 0E\exp\left(iN_T^n + \frac{1}{2}<N>_T\right) \to 1. \quad (4.10)$$

Thus, it is sufficient to prove

$$e(A^n)_T \to e^{-\frac{1}{2}<N>_T}.$$

Recall

$$A_t^n = -\frac{1}{2} < N >_t + \int_0^t \int_{|x|\leq d_n} (e^{ix} - 1 - ix)\tilde{\nu}^n(ds, dx).$$

Let

$$\alpha_t^n = \int_{|x|\leq d_n} (e^{ix} - 1 - ix)\tilde{\nu}^n(\{t\}, dx).$$

Since $(e^{ix} - 1 - ix) \leq x^2/2$, we have $\alpha_t^n \leq d_n^2/2$. Therefore,

$$\begin{aligned}\sum_{0\leq t\leq T} |\alpha_t^n| &= \frac{1}{2}\int_0^T \int_{|x|\leq d_n} x^2\tilde{\nu}^n(dt, dx)\\ &= \frac{1}{2} < N^n >_T\\ &= \frac{1}{2}\frac{\lambda^2 a}{2}.\end{aligned}$$

Then,

$$\prod_{0<t\leq T} (1 + \alpha_t^n)e^{-\alpha_t^n} \to 1.$$

By definition of $e(A)_t$, it remains to prove

$$\frac{1}{2} < N^{nc} >_T - \int_0^T \int_{|x|\leq d_n} (e^{isx} - 1 - isx)\tilde{\nu}^n(ds, dx) \to_p \frac{1}{2} < N >_T.$$

By observation (a) and the form of $< N^n >_T$, it suffices to prove

$$\int_0^T \int_{|x|\leq d_n} (e^{isx} - 1 - isx)\tilde{\nu}^n(ds, dx) \to_p 0.$$

We have, since $|(e^{isx} - 1 - isx + x^2/2)| \le c_s \frac{x^3}{3}$,

$$\int_0^T \int_{|x|\le d_n} \underbrace{\left((e^{isx} - 1 - isx) + \frac{x^2}{2}\right)}_{\le \frac{|x|^3}{6}} \tilde{v}^n(ds, dx) \le \frac{d_n}{6} \int_0^T \int_{|x|\le d_n} x^2 \tilde{v}^n(ds, dx)$$
$$\le \frac{d_n}{6} < N^n >_T$$
$$\le \frac{d_n}{6} \lambda^2 a$$
$$\longrightarrow 0.$$

To dispose of assumption, define

$$\tau_n = \min\{t \le T :< M^n >_t \ge < M >_T +1\}.$$

Then τ_n is stopping time. We have $\tau_n = T$ if $< M^n >_t << M >_T +1$. Let $\tilde{M}^n = M^n_{t\wedge\tau_n}$. Then

$$< \tilde{M}^n >_T \le 1+ < M >_T +\epsilon_n^2 \le 1+ < M >_T +\epsilon_1^2$$

and

$$\lim_n P\left(| < \tilde{M}^n >_t - < \tilde{M} >_t | > \epsilon\right) \le \lim_n P(\tau_n > T) = 0.$$

Next,

$$\lim_{n\to\infty} Ee^{iN_T^n} = \lim_{n\to\infty} E\left(e^{iN_t^n} - e^{iN^n_{t\wedge\tau_n}}\right) + \lim_{n\to\infty} Ee^{iN^n_{t\wedge\tau_n}}$$
$$= \lim_{n\to\infty} E\left(e^{iN_T^n} - e^{iN^n_{T\wedge\tau_n}}\right) + Ee^{iN_T}$$
$$= Ee^{iN_T}.$$

The last equality follows from

$$\lim_{n\to\infty} \left|E\left(e^{iN_T^n} - e^{iN^n_{T\wedge\tau_n}}\right)\right| \le 2 \lim_{n\to\infty} P(\tau_n > T) = 0.$$

This completes the proof.

4.4 Application to survival analysis

Let X be a positive random variable. Let F and f be cumulative distribution function and probability density function of X. Then, survival function $\bar{F}$ is defined as

$$\bar{F}(t) = P(X > t) = 1 - F(t)$$

Then, we have

$$\begin{aligned} P(t < X \le t + \triangle t | X > t) &= \frac{P(t < X \le t + \triangle t)}{\bar{F}(t)} \\ &= \frac{\int_t^{t+\triangle t} dF(s)}{\bar{F}(t)}. \end{aligned}$$

Since we know

$$\frac{1}{\triangle t} \int_t^{t+\triangle t} f(x) ds \longrightarrow f(t).$$

as $\triangle t \to 0$, hazard rate, is defined as

$$\begin{aligned} h(t) &= \frac{f(t)}{\bar{F}(t)} \\ &= -\frac{d}{dt} \log \bar{F}(t). \end{aligned}$$

Therefore, survival function can be written as

$$\bar{F}(t) = \exp\left(- \int_0^t h(s) ds \right).$$

If the integrated hazard rate is given, then it determines uniquely life distribution. For example, think about the following:

$$\tau_F = \sup\{s : F(s) < 1\}.$$

Consider now the following problem arising in clinical trials.
Let

1. $X_1, \ldots., X_n$ be i.i.d F (life time distribution)
2. $U_1, \ldots., U_n$ be i.i.d measurable function with distribution function G with $G(\infty) < 1$, which means U_i are not random variable in some sense.

Now, consider

1. indicator for "alive or not at time s": $1(X_i \le U_i, X_i \wedge U_i \le s)$.
2. indicator for "alive and leave or not at time s": $U_i 1(X_i \wedge U_i \le s)$.
3. indicator for "leave or not at time s": $1(X_i \wedge U_i \ge s)$.

and σ-field

$$\mathcal{F}_t^n = \sigma\Big(\{1(X_i \le U_i, X_i \wedge U_i \le s), U_i 1(X_i \wedge U_i \le s), 1(X_i \wedge U_i \ge s), s \le t, i = 1, 2, \ldots, n\}\Big).$$

$\mathcal{F}_t^n$ is called information contained in censored data.

Let

$$\beta(t) = \int_0^t h(s)ds.$$

If $\hat{\beta}(t)$ is an estimate of $\beta(t)$, then we can estimate survival function. The estimator of survival function will be

$$\hat{\bar{F}}(t) = e^{-\hat{\beta}(t)},$$

which will be approximately be

$$\prod_{s\le t}\left(1 - d(\hat{\beta}(s))\right).$$

This is alternate estimate of survival function, which is called, Nelson estimate.

Let

$$N_n(t) = \sum_{i=1}^{n} 1(X_i \le U_i, X_i \wedge U_i \le t)$$

$$Y_n(t) = \sum_{i=1}^{n} 1(X_i \wedge U_i \ge t).$$

Then, $\hat{\beta}(t)$, which is called Breslow estimator, will be

$$\hat{\beta}(t) = \int_0^t \frac{dN_n(s)}{Y_n(s)} \approx \frac{\int_t^{t+\Delta t} dF(s)}{\bar{F}(t)}.$$

Now, we consider another estimator of survival function, which is called the Kaplan-Meier estimator. It will be

$$\prod_{s\le t}\left(1 - \frac{\Delta N_n(s)}{Y_n(s)}\right)$$

Gill [12] showed the asymptomatic properties of the Kaplan-Meier estimator.

We can show that

$$\left|e^{-\hat{\beta}(t)} - \prod_{s\le t}\left(1 - \frac{\Delta N_n(s)}{Y_n(s)}\right)\right| = O\left(\frac{1}{n}\right) \tag{4.11}$$

using following lemma.

Lemma 4.5: Let $\{\alpha^n(s), 0 \le s \le T, n \ge 1\}$ be real-valued function such that
1. $\{s \in (0, u] : \alpha^n(s) \neq 0\}$ is P-a.e. at most countable for each n.
2. $\sum_{0<s\le u} |\alpha^n(s)| \le C$ with C constant.
3. $\sup_{s\le u}\{|\alpha^n(s)|\} = O(a_n)$, where $a_n \searrow 0$ as n goes ∞.

Then,

$$\sup_{t\le u}\left|\prod_{0<s\le t}(1-\alpha^n(s))-\prod_{0<s\le t}e^{-\alpha^n(s)}\right| = O(a_n).$$

Proof: We choose n_0 large such that for $n \ge n_0$ $O(a_n) < \frac{1}{2}$. Since

$$\begin{aligned}\prod_{0<s\le t}(1-\alpha^n(s))e^{\alpha^n(s)} &= \exp\Big(\sum_{0<s\le t}\log(1-\alpha^n(s))+\alpha^n(s)\Big)\\ &= \exp\Big(\sum_{0<s\le t}\sum_{j=2}^{\infty}\frac{(-1)^{j+2}}{j}(\alpha^n(s))^j\Big) \text{ (by Taylor expansion.)},\end{aligned}$$

for $n \ge n_0$, we have

$$\begin{aligned}\left|\prod_{0<s\le t}(1-\alpha^n(s))e^{\alpha^n(s)}-1\right| &\le \left|\exp\Big(\sum_{0<s\le t}\sum_{j=2}^{\infty}\frac{(-1)^{j+2}}{j}(\alpha^n(s)^j-1)\Big)\right|\\ &\le e^{\eta}\left|\sum_{0<s\le t}\sum_{j=2}^{\infty}\frac{(-1)^{j+2}}{j}(\alpha^n(s))^j\right|.\end{aligned}$$

where

$$0\wedge\sum_{0<s\le t}\sum_{j=2}^{\infty}\frac{(-1)^{j+2}}{j}(\alpha^n(s))^j < \eta < 0\vee\sum_{0<s\le t}\sum_{j=2}^{\infty}\frac{(-1)^{j+2}}{j}(\alpha^n(s))^j.$$

For large n,

$$\begin{aligned}\left|\sum_{0<s\le t}\sum_{j=2}^{\infty}\frac{(-1)^{j+2}}{j}(\alpha^n(s))^j\right| &\le \sum_{0<s\le t}\sum_{j=2}^{\infty}\frac{|\alpha^n(s)|^j}{j}\\ &\le \underbrace{\sup_{s\le u}|\alpha^n(s)|}_{=O(a_n)}\ \underbrace{\sum_{0<s\le t}|\alpha^n(s)|}_{\le t\cdot M}\ \underbrace{\sum_{j=1}^{\infty}\Big(\frac{1}{2}\Big)^{j-2}\frac{1}{j}}_{<\infty}\\ &\longrightarrow 0.\end{aligned}$$

$\sum_{0<s\le t}|\alpha^n(s)| \le t\cdot M$ holds since $|\alpha^n(s)|$ will be bounded by M.

To prove (4.11), we let

$$\alpha^n(s) = \frac{1}{Y_n(s)}, \quad s \le T$$

and

$$\Delta N(s) = 0, \quad s > T.$$

We get $a_n = 1/n$ using the Glivenko-Cantelli theorem.

4.5 Asymptotic distribution of $\hat{\beta}(t)$ and Kaplan-Meier estimate

X_i and U_i are defined as previously. Again, define $N_n(t)$, $Y_n(t)$ as

$$\begin{aligned} N_n(t) &= \sum_{i=1}^{n} 1(X_i \le U_i, X_i \wedge U_i \le t) \\ Y_n(t) &= \sum_{i=1}^{n} 1(X_i \wedge U_i \ge t). \end{aligned}$$

Then,

$$\begin{aligned} \beta(t) &= \int_0^t h(s)ds \\ &= \int_0^t \frac{f(s)}{\bar{F}(s)} ds \\ &= \int_0^t \frac{dF(s)}{1-F(s)} \\ \hat{\beta}(t) &= \int_0^t \frac{dN_n(s)}{Y_n(s)}. \end{aligned}$$

Using the above lemma, we can show that the difference between Kaplan-Meier and Nelson estimates is asymptotically of order $\frac{1}{n}$.

$$\begin{aligned} &\sup_{t\le u}\left|\prod_{s\le t}\left(1-\frac{\triangle N_n(s)}{Y_n(s)}\right)-e^{-\hat{\beta}(t)}\right| \\ &= \sup_{t\le u}\left|\prod_{s\le t}\left(1-\frac{\triangle N_n(s)}{Y_n(s)}\right)-\exp\left(-\int_0^t \frac{dN_n(s)}{Y_n(s)}\right)\right| \\ &= O_P\left(\frac{1}{n}\right). \end{aligned}$$

Let

$$\begin{aligned} Q_1(s) &= P(X_1 \wedge U_1 \le s, X_1 \le U_1) \\ H(s) &= P(X_1 \wedge U_1 \le s) \\ \beta_1(t) &= \int_0^t \frac{dQ_1(s)}{(1-H(s-))}. \end{aligned}$$

Assume that X (with F) and U (with G) are independent. Then,

$$\begin{aligned} \triangle Q_1(s) &= P(s \le X_1 \wedge U_1 \le s+\triangle s, X_1 \le U_1) \\ &= P(s \le X_1 \le s+\triangle s, X_1 \le U_1) \\ &= P(s \le X_1 \le s+\triangle s, U_1 \ge s+\triangle s) \\ &= P(s \le X_1 \le s+\triangle s)\cdot P(U_1 \ge s+\triangle s) \\ dQ_1(s) &= (1-G(s-))dF(s). \end{aligned}$$

Similarly,

$$(1 - H(s-)) = (1 - G(s-))(1 - F(s)). \tag{4.12}$$

Then,

$$\begin{aligned}\beta_1(t) &= \int_0^t \frac{dQ_1(s)}{(1 - H(s-))}\\ &= \int_0^t \frac{(1 - G(s-))dF(s)}{(1 - G(s-))(1 - F(s))}\\ &= \int_0^t \frac{dF(s)}{(1 - F(s))}.\end{aligned}$$

Lemma 4.6 (PL 1): Let

$$\mathcal{F}_t^n = \sigma\Big(\{1(X_i \le U_i, X_i \wedge U_i \le s), U_i 1(X_i \wedge U_i \le s), 1(X_i \wedge U_i \ge s), s \le t, i = 1, 2, \ldots, n\}\Big).$$

Suppose we have

1. $(N_n(t), \mathcal{F}_t^n)$ is a point process and

   $$\left\{N_n(t) - \int_0^t Y_n(s)\frac{dQ_1(s)}{(1 - H(s-))}\right\}$$

   is martingale.
   and
2. $\hat{\beta}_n(t) = \int_0^t \frac{dN_n(t)}{Y_n(t)}$.

Then, $m_n(t) = \hat{\beta}_n(t) - \beta_1(t)$ is a locally square integrable martingale, and increasing process $< m_n >_t$ will be

$$< m_n >_t = \int_0^t \frac{1}{Y_n(s)} \frac{dQ_1(s)}{(1 - H(s-))}$$

Remark: If X and U are independent, from (4.12), we have

$$\frac{dQ_1(s)}{(1 - H(s-))} = \frac{dF(s)}{1 - F(s)}$$

Let us assume that $\{X_i\}$ and $\{U_i\}$ are independent and

$$A_n(t) = \int_0^t Y_n(s)\frac{dQ_1(s)}{(1 - H(s-))}.$$

From the previous theorem we know that $m_n(t) = (\hat{\beta}_n(t) - \beta_1(t))$ is a locally square integrable martingale with

$$< m_n >_t = \int_0^t \frac{1}{Y_n(s)} \frac{dQ_1(s)}{(1 - H(s-))}.$$

Hence

$$\sqrt{n}(\hat{\beta}_n(t) - \beta_1(t)) \Rightarrow_{a.s.} \gamma_t$$

where Y_t is a Gaussian martingale

$$\begin{aligned}
\langle \sqrt{n} m_n \rangle_t &= \int_0^t \frac{n}{Y_n(s)} \frac{dQ_1(s)}{(1 - H(s-))} \to C_1(t) \\
C_1(t) &= \int_0^t \frac{dQ_1(s)}{(1 - H(s-))^2} \\
r_t &= C_1(t).
\end{aligned}$$

Also $\langle m_n \rangle_t = A_n(t)$ which gives by Glivenko-Cantelli Lemma $\langle m_n \rangle_t = 0_p\left(\frac{1}{n}\right)$. Using Lenglart inequality we get

$$\sup_{s \le t} \left| \hat{\beta}_n(s) - \beta_1(s) \right| \to_P 0. \tag{4.13}$$

Hence $\hat{\beta}_n(s)$ is consistent estimate of integrated hazard rate under the independence assumption above. We note that under the assumption

$$\beta_1(t) = \int_0^t \frac{dF(s)}{(1 - F(s))}$$

and

$$C_1(t) = \int_0^1 \frac{dF(s)}{(1 - F(s))^2 (1 - G(s-))}.$$

With $\tau_H = \inf\{s : H(s-) < 1\}$, the above results hold for all $t < \tau_H$ only.

Lemma 4.7: 1. For $t < \tau_H$

$$\sup_{s \le t} \left| \exp\left(- \int_0^t \frac{dF(s)}{(1 - F(s))} \right) - \exp\left(- \int_0^t \frac{dN_n(s)}{Y_n(s)} \right) \right| \to_p 0.$$

2.

$$\sqrt{n} \left[\exp\left(- \int_0^{\cdot} \frac{dF(s)}{(1 - F(s))} \right) - \exp\left(- \int_0^{\cdot} \frac{dN_n(s)}{Y_n(s)} \right) \right] \to_{\infty} r \cdot \exp\left(- \int_0^{\cdot} \frac{dF(s)}{(1 - F(s))} \right)$$

in $D(0, t]$ for $t < \tau_H$ with γ as above.

Proof: Using Taylor expansion we get

$$\exp\left(-\int_0^t \frac{dF(s)}{(1-F(s))}\right) - \exp(-\hat{\beta}_n(t))$$
$$= \exp\left(-\int_0^t \frac{dF(s)}{(1-F(s))}\right)\left\{(\beta_1(t) - \hat{\beta}_n(t)) + \frac{(\beta_1(t) - \hat{\beta}_n(t))^2}{2}\exp(-h_n)\right\}$$

with h_n is a random variable satisfying $\beta_1(t) \wedge \hat{\beta}_n(t) \le h_n \le \beta_1(t) \vee \hat{\beta}_n(t)$. Since for $t < \tau_H$, $\exp(-h_n)$ is bounded by convergence of $\sup_{s\le t}(\hat{\beta}_n(s) - \beta_1(s)) \to_p 0$, the result follows. To prove the second part, note that

$$\sqrt{n}(\beta_1(\cdot) - \hat{\beta}_n(\cdot))^2 = \sqrt{n}(\beta_1(\cdot) - \hat{\beta}_n(\cdot))(\beta_1(\cdot) - \hat{\beta}_n(\cdot)) \Rightarrow_{\mathcal{D}} \gamma \cdot 0 = 0$$

by Slutsky theorem and the first term converges in distribution to γ.

Theorem 4.3 (R. Gill): Let $\hat{F}_n(t) = \prod_{s\le t}\left(1 - \frac{\Delta N_n(s)}{Y_n(s)}\right)$. Then under independence of $\{X_i\}$, $\{U_i\}$, we get that

$$\frac{n(\hat{F}_n(\cdot) - F(\cdot))}{1 - F(\cdot)} \Rightarrow \gamma_o$$

in $D[0, t]$ for $t < \tau_H$.

Proof: We have $\exp(-\beta_1(t)) = 1 - F(t)$ for $t < \tau_H$. Hence by previous Lemma

$$\frac{\sqrt{n}}{(1 - F(\cdot))}(\exp(-\hat{\beta}_n(\cdot)) - (1 - F(\cdot))) \Rightarrow_{\mathcal{D}} \gamma.$$

Although we do not present it here, similar techniques work to study so-called Linden-Bell estimator which arises in the study of truncated data (e.g. in randomly starting clinical trials). The latter has applications in astronomy. Interested readers are referred to ([26], [21]).

5 Central limit theorems for dependent random variables

When one collects data, the observed sample could produce dependent random variables. A simplest example of this could be

$$S_{n,m} = \sum_{1}^{m} X_{n,m},$$

which could be a martingale for each n. These kind of theorems were first considered by Billingsley [3]. A major breakthrough for convergence of interpolated $S_{n,[n,t]}$ was obtained by Dvoretsky [10]. Gordin [13] showed that for second-order stationary processes with "mixing" condition, the central limit problem can be reduced to that for martingales. We show, following Durrett and Resnick [9], that one can derive the convergence of the interpolated sequence associated with the martingale central limit theorem, which can be obtained by their extension of the Skorokhod embedding theorem and using the work in [8] to obtain convergence in the of the interpolated sequence associated with the stationary sequence central limit problem of the Brownian motion. This proves weak convergence of such sequence in $C[0, 1]$.

In this chapter, we study the central limit theorems for dependent random variables using the Skorokhod embedding theorem.

Theorem 5.1 (Martingale central limit theorem (discrete): Let $\{S_n\}$ be a martingale. Let $S_0 = 0$ and $\{W(t), 0 \le t < \infty\}$ be Brownian motion. Then there exists a sequence of stopping time, $0 = T_0 \le T_1 \le T_2 \cdots \le T_n$, with respect to $\mathcal{F}_t^W$ such that

$$(S_0, \ldots, S_n) =_d (W(T_0), \ldots, W(T_n)).$$

Proof: We use induction.

$$T_0 = 0.$$

Assume there exists $(T_0, \ldots, T_{k-1})$ such that

$$(S_0, \ldots, S_{k-1}) =_d (W(T_0), \ldots, W(T_{k-1})).$$

Note that the strong Markov property implies that $\{W(T_{k-1} + t) - W(T_{k-1}), t \ge 0\}$ is a Brownian motion, independent of $\mathcal{F}_t^W$. Look at the regular conditional distribution of $S_k - S_{k-1}$ given $S_0 = s_0, \ldots, S_{k-1} = s_{k-1}$. Denote it by

$$\mu(S_0, \ldots, S_{k-1}; B) = P\Big(S_k - S_{k-1} \in B \big| S_0 = S_0, \ldots, S_{k-1} = S_{k-1}\Big) \text{ for } B \in (B(\mathbb{R})).$$

$$\text{So } \mu(S_0, S_1, \ldots, S_{k-1}; B) = P\Big(S_k - S_{k-1} \in B \big| S_0 \ldots, S_{k-1}\Big).$$

Since S_k is a martingale, we have

$$0 = E\left(S_k - S_{k-1} | S_0, \dots, S_{k-1}\right) = \int x\mu_k(S_0, \dots, S_{k-1}; dx).$$

By Skorokhod's representation theorem, we see that for a.e. $S \equiv (S_0, \dots, S_{k-1})$, there exists a stopping time $\tilde{\tau}_S$(exist time from (U_k, V_k)) such that

$$W(T_{k-1} + \tilde{\tau}_S) - W(T_{k-1}) = \tilde{W}(\tau_k) =_d \mu_k(S_0, \dots, S_{k-1}; \cdot).$$

We let $T_k = T_{k-1} + \tilde{\tau}_S$, then

$$(S_0, S_1, \dots, S_k) =_d (W(T_0), \dots, W(T_k)),$$

and the result follows by induction.

Remark: If $E(S_k - S_{k-1})^2 < \infty$, then

$$E\left(\tilde{\tau}_S | S_0, \dots, S_{k-1}\right) = \int x^2 \mu_k(S_0, \dots, S_{k-1}; dx)$$

since $W_t^2 - t$ is a martingale and $\tilde{\tau}_S$ is the exit time from a randomly chosen interval $(S_{k-1} + U_k, S_{k-1} + V_k)$.

Definition 5.1: We say that $X_{n,m}, \mathcal{F}_{n,m}, 1 \le m \le n$, is a martingale difference array if $X_{n,m}$ is $\mathcal{F}_{n,m}$ measurable and $E(X_{n,m}|\mathcal{F}_{n,m-1}) = 0$ for $1 \le m \le n$, where $\mathcal{F}_{n,0} = \{\emptyset, \Omega\}$.

Notation: Let

$$S_{(u)} = \begin{cases} S_k, & \text{if } u = k \in \mathbf{N}; \\ \text{linear on } u, & \text{if } u \in [k, k+1] \text{ for } k \in \mathbf{N}. \end{cases}$$

and

$$S_{n,(u)} = \begin{cases} S_{n,k}, & \text{if } u = k \in N; \\ \text{linear on } u, & \text{if } u \in [k, k+1]. \end{cases}$$

Consider $X_{n,m} (1 \le m \le n)$ be triangular arrays of random variables with

$$\begin{aligned} EX_{n,m} &= 0 \\ S_{n,m} &= X_{n,1} + \dots + X_{n,m}. \end{aligned}$$

We shall use the following fact:

$$S_{n,m} = W(\tau_m^n)$$

and

$$\tau^n_{[ns]} \to_P s \text{ for } s \in [0,1],$$

then $||S_{n,(n\cdot)} - W(\cdot)||_\infty \to_P 0$.

Theorem 5.2: Let $\{X_{n,m}, \mathcal{F}_{n,m}\}$ be a martingale difference array and $S_{n,m} = X_{n,1} + \cdots + X_{n,m}$. Assume that

1. $|X_{n,m}| \le \epsilon_n$ for all m, and $\epsilon_n \to 0$ as $n \to \infty$
2. with $V_{n,m} = \sum_{k=1}^m E\left(X^2_{n,k}|\mathcal{F}_{n,k-1}\right)$, $V_{n,[nt]} \to t$ for all t.

Then $S_{n,(n\cdot)} \Rightarrow W(\cdot)$.

Proof: We stop $V_{n,k}$ first time if it exceeds 2 (call it k_0) and set $X_{n,m} = 0$, $m > k_0$. We can assume without loss of generality

$$V_{n,n} \le 2 + \epsilon_n^2$$

for all n. Using Theorem 5.1, we can find stopping times $T_{n,1}, \ldots, T_{n,n}$ so that

$$(0, S_{n,1}, \ldots, S_{n,n}) =_{\mathcal{D}} (W(0), W(T_{n,1}), \ldots, W(T_{n,n})).$$

Using Lemma 5.1 and the above equality, it suffices to show that $T_{n,[nt]} \to_P t$ for each t. Let

$$t_{n,m} = T_{n,m} - T_{n,m-1} \text{ with } (T_{n,0} = 0).$$

Using the Skohorod embedding theorem, we have

$$E\left(t_{n,m}|\mathcal{F}_{n,m-1}\right) = E\left(X^2_{n,m}|\mathcal{F}_{n,m-1}\right).$$

The last observation with Hypothesis 2 implies

$$\sum_{m=0}^{[nt]} E\left(t_{n,m}|\mathcal{F}_{n,m-1}\right) \to_P t.$$

Observe that

$$\begin{aligned} E\left(T_{n,[nt]} - V_{n,[nt]}\right)^2 &= E\left(\sum_{m=1}^{[nt]} \underbrace{\left(t_{n,m} - E\left(t_{n,m}|\mathcal{F}_{n,m-1}\right)\right)}_{\text{any two terms are orthogonal}}\right)^2 \\ &= \sum_{m=1}^{[nt]} E\left(t_{n,m} - E\left(t_{n,m}|\mathcal{F}_{n,m-1}\right)\right)^2 \end{aligned}$$

$$\begin{aligned}
&\le \sum_{m=1}^{[nt]} E\Big(t_{n,m}^2 | \mathcal{F}_{n,m-1}\Big) \\
&\le \sum_{m=1}^{[nt]} C\cdot E\Big(X_{n,m}^4 | \mathcal{F}_{n,m-1}\Big) \text{ (we will show} \\
&\quad \text{that } C = 4.) \\
&\le \sum_{m=1}^{[nt]} C\epsilon_n^2 E\Big(X_{n,m}^2 | \mathcal{F}_{n,m-1}\Big) \text{ (by Assumption 1)} \\
&= C\epsilon_n^2 V_{n,n} \\
&\le C\epsilon_n^2 (2+\epsilon_n^2) \to 0. \qquad (5.1)
\end{aligned}$$

Since L^2 convergence implies convergence in probability,

$$E\Big(T_{n,[nt]} - V_{n,[nt]}\Big)^2 \longrightarrow 0$$

and

$$V_{n,[nt]} \longrightarrow {}_P t$$

together implies

$$T_{n,[nt]} \longrightarrow {}_P t.$$

Proof of (5.1): If θ is real, then

$$E\Big(\exp\Big(\theta(W(t)-W(s)) - \frac{1}{2}\theta^2(t-s)\Big)\Big|\mathcal{F}_s^W\Big) = 1.$$

Since

$$E\Big(\exp\Big(\theta W(t) - \frac{1}{2}\theta^2 t\Big)\Big|\mathcal{F}_s^W\Big) = \exp\Big(\theta W(s) - \frac{1}{2}\theta^2 s\Big),$$

we know that $\left\{\exp\left(\theta W(t) - \frac{1}{2}\theta^2 t\right), \mathcal{F}_t^W\right\}$ is a martingale. Then, for all $A \in \mathcal{F}_s^W$,

$$\begin{aligned}
E1_A\Big(\exp\Big(\theta W(s) - \frac{1}{2}\theta^2 s\Big)\Big) &= \int_A \Big(\exp\Big(\theta W(s) - \frac{1}{2}\theta^2 s\Big)\Big) dP \\
&= \int_A \exp\Big(\theta W(t) - \frac{1}{2}\theta^2 t\Big) dP \\
&\quad \text{(by definition of conditional expectation)} \\
&= E1_A\Big(\exp\Big(\theta W(t) - \frac{1}{2}\theta^2 t\Big)\Big).
\end{aligned}$$

Take a derivative in θ and find a value at $\theta = 0$.

Number of derivative		
1	$W(t)$ is MG	
2	$W^2(t) - t$ is MG	(5.2)
3	$W^3(t) - 3tW(t)$ is MG	
4	$W^4(t) - 6tW^2(t) + 3t^2$ is MG	

For any stopping time τ,

$$E(W^4(\tau) - 6\tau W^2(\tau) + 3\tau^2) = 0.$$

Therefore,

$$\begin{aligned} &EW_\tau^4 - 6E(\tau W_\tau^2) = -3EW_\tau^2 \\ \Rightarrow\ &EW_\tau^2 \leq 2E(\tau W_\tau^2). \end{aligned}$$

Since

$$E(\tau W_\tau^2) \leq \left(E\tau^2\right)^{1/2} \cdot \left(EW_\tau^4\right)^{1/2}$$

by Schwartz Inequality, we have

$$\left(E\tau^2\right)^{1/2} \leq 2\left(EW_\tau^4\right)^{1/2}.$$

Therefore,

$$E\left(t_{n,m}^2 | \mathcal{F}_{n,m-1}\right) \leq 4E\left(X_{n,m}^4 | \mathcal{F}_{n,m-1}\right).$$

Theorem 5.3 Generalization of the Lindberg-Feller theorem: Let $\{X_{n,m}, \mathcal{F}_{n,m}\}$ be a martingale difference array and $S_{n,m} = X_{n,1} + \cdots + X_{n,m}$. Assume that

1. $V_{n,[nt]} = \sum_{k=1}^{[nt]} E\left(X_{n,k}^2 | \mathcal{F}_{n,k-1}\right) \to_P t$.
2. $\widehat{V}(\epsilon) = \sum_{m \leq n} E\left(X_{n,m}^2 1(|X_{n,m}| > \epsilon) | \mathcal{F}_{n,m-1}\right) \to_P 0$, for all $\epsilon > 0$.

Then $S_{n,(n\cdot)} \Rightarrow W(\cdot)$.

For i.i.d. $X_{n,m}$ and $t = 1$, we get the Lindberg-Feller theorem.

Lemma 5.2: There exists $\epsilon_n \to 0$ such that $\epsilon_n^2 \hat{V}(\epsilon_n) \to_P 0$.

Proof: Since $\hat{V}(\epsilon) \to_P 0$, we choose large N_m such that

$$P\left(\hat{V}\left(\frac{1}{m}\right) > \frac{1}{m^3}\right) < \frac{1}{m}$$

for $m \geq N_m$. Here we choose $\epsilon_n = \frac{1}{m}$ with $n \in [N_m, N_{m+1})$. For $\delta > \frac{1}{m}$, we have

$$P\left(\epsilon_n^{-2}\widehat{V}(\epsilon_n) > \delta\right) \leq P\left(m^2\widehat{V}\left(\frac{1}{m}\right) > \frac{1}{m}\right) < \frac{1}{m}.$$

This completes the proof of lemma.

Let

$$\begin{aligned}
\overline{X}_{n,m} &= X_{n,m}1(|X_{n,m}| \leq \epsilon_n) \\
\widehat{X}_{n,m} &= X_{n,m}1(|X_{n,m}| > \epsilon_n) \\
\widetilde{X}_{n,m} &= \overline{X}_{n,m} - E\left(\overline{X}_{n,m}|\mathcal{F}_{n,m-1}\right).
\end{aligned}$$

Lemma 5.3: $\widetilde{S}_{n,[n\cdot]} \Rightarrow W(\cdot)$

Proof: We will show that $\widetilde{X}_{n,m}$ satisfies Theorem 5.2.

$$\begin{aligned}
|\widetilde{X}_{n,m}| &= \left|\overline{X}_{n,m} - E\left(\overline{X}_{n,m}|\mathcal{F}_{n,m-1}\right)\right| \\
&\leq \left|\overline{X}_{n,m}\right| + \left|E\left(\overline{X}_{n,m}|\mathcal{F}_{n,m-1}\right)\right| \\
&\leq 2\epsilon_n \to 0.
\end{aligned}$$

and hence, the first condition is satisfied. Since

$$X_{n,m} = \overline{X}_{n,m} + \widehat{X}_{n,m},$$

we have

$$\begin{aligned}
E\left(\overline{X}^2_{n,m}|\mathcal{F}_{n,m-1}\right) &= E\left((X_{n,m} - \widehat{X}_{n,m})^2|\mathcal{F}_{n,m-1}\right) \\
&= E\left(X^2_{n,m} - 2X_{n,m}\widehat{X}_{n,m} + \widehat{X}_{n,m})^2|\mathcal{F}_{n,m-1}\right) \\
&= E\left(X^2_{n,m}|\mathcal{F}_{n,m-1}\right) - E\left(\widehat{X}^2_{n,m}|\mathcal{F}_{n,m-1}\right). \qquad (5.3)
\end{aligned}$$

The last equality follows from $E\left(X_{n,m}\widehat{X}_{n,m}|\mathcal{F}_{n,m-1}\right) = E\left(\widehat{X}^2_{n,m}|\mathcal{F}_{n,m-1}\right)$. Since $X_{n,m}$ is a martingale difference array, and hence, $E(X_{n,m}|\mathcal{F}_{n,m-1}) = 0$. The last observation implies $E\left(\overline{X}_{n,m}|\mathcal{F}_{n,m-1}\right) = -E\left(\widehat{X}_{n,m}|\mathcal{F}_{n,m-1}\right)$, and hence,

$$\begin{aligned}
\left[E\left(\overline{X}_{n,m}|\mathcal{F}_{n,m-1}\right)\right]^2 &= \left[E\left(\widehat{X}_{n,m}|\mathcal{F}_{n,m-1}\right)\right]^2 \\
&\leq E\left(\widehat{X}^2_{n,m}|\mathcal{F}_{n,m-1}\right) \text{ (by Jensen's inequality).}
\end{aligned}$$

Therefore,

$$\begin{aligned}\sum_{m=1}^{n}\left[E\left(\overline{X}_{n,m}|\mathcal{F}_{n,m-1}\right)\right]^2 &\le \sum_{m=1}^{n}E\left(\widehat{X}^2_{n,m}|\mathcal{F}_{n,m-1}\right)\\ &= \widehat{V}(\epsilon_n)\\ &\to_P 0\end{aligned}$$

by given condition. Finally,

$$\begin{aligned}\sum_{m=1}^{n}E\left(\widetilde{X}^2_{n,m}|\mathcal{F}_{n,m-1}\right) &= \sum_{m=1}^{n}E\left(\overline{X}^2_{n,m}|\mathcal{F}_{n,m-1}\right)\\ &\quad-\sum_{m=1}^{n}E\left(\overline{X}_{n,m}|\mathcal{F}_{n,m-1}\right)^2\\ &\quad\text{(by the conditional variance formula)}\\ &= \sum_{m=1}^{n}\left(E\left(X^2_{n,m}|\mathcal{F}_{n,m-1}\right)-E\left(\widehat{X}^2_{n,m}|\mathcal{F}_{n,m-1}\right)\right)\\ &\quad-\sum_{m=1}^{n}E\left(\overline{X}_{n,m}|\mathcal{F}_{n,m-1}\right)^2\\ &\quad\text{(from equation (5.3))}\\ &= \sum_{m=1}^{n}E\left(X^2_{n,m}|\mathcal{F}_{n,m-1}\right)-\sum_{m=1}^{n}E\left(\widehat{X}^2_{n,m}|\mathcal{F}_{n,m-1}\right)\\ &\quad-\sum_{m=1}^{n}\left[E\left(\overline{X}_{n,m}|\mathcal{F}_{n,m-1}\right)\right]^2.\end{aligned}$$

As $\hat{V}(\epsilon_n)\xrightarrow{p}0$ and the last term converges to zero in probability, we get

$$\lim_{n\to\infty}\sum_{m=1}^{n}E\left(\widetilde{X}^2_{n,m}|\mathcal{F}_{n,m-1}\right)=\lim_{n\to\infty}\sum_{m=1}^{n}E\left(X^2_{n,m}|\mathcal{F}_{n,m-1}\right).$$

Since

$$V_{n,[nt]}=\sum_{m=1}^{[nt]}E\left(X^2_{n,m}|\mathcal{F}_{n,m-1}\right)\to_P t,$$

we conclude that

$$\sum_{m=1}^{[nt]}E\left(\widetilde{X}^2_{n,m}|\mathcal{F}_{n,m-1}\right)\to_P t,$$

This show that the second condition is satisfied, and this completes the proof.

Lemma 5.4:

$$||S_{n,(n\cdot)} - \widetilde{S}_{n,(n\cdot)}||_\infty \leq \sum_{m=1}^{n} \left| E\left(\overline{X}_{n,m} | \mathcal{F}_{n,m-1}\right)\right|.$$

Proof: Note that if we prove this lemma, then, since $\sum_{m=1}^{n} \left|E\left(\overline{X}_{n,m}|\mathcal{F}_{n,m-1}\right)\right| \to_P 0$(we will show this), and we construct a Brownian motion with $||\widetilde{S}_{n,(n\cdot)} - W(\cdot)||_\infty \to 0$, the desired result follows from the triangle inequality.

Since $X_{n,m}$ is martingale difference array, we know that $E\left(\overline{X}_{n,m}|\mathcal{F}_{n,m-1}\right) = -E(\widehat{X}_{n,m}|\mathcal{F}_{n,m-1})$, and hence,

$$\begin{aligned}
\sum_{m=1}^{n} \left|E\left(\overline{X}_{n,m}|\mathcal{F}_{n,m-1}\right)\right| &= \sum_{m=1}^{n} \left|E\left(\widehat{X}_{n,m}|\mathcal{F}_{n,m-1}\right)\right| \\
&\leq \sum_{m=1}^{n} E\left(|\widehat{X}_{n,m}||\mathcal{F}_{n,m-1}\right) \text{ by Jensen} \\
&\leq \frac{1}{\epsilon_n} \sum_{m=1}^{n} E\left(\widehat{X}_{n,m}^2|\mathcal{F}_{n,m-1}\right) \\
&\quad \left(\text{if } |X_{n,m}| > \epsilon_n, \widehat{X}_{n,m} \leq \frac{X_{n,m}^2}{\epsilon_n} = \frac{\widehat{X}_{n,m}^2}{\epsilon_n}\right) \\
&= \frac{\widehat{V}(\epsilon_n)}{\epsilon_n} \to_p 0 \text{ (by Lemma 5.2).}
\end{aligned}$$

On $\{|X_{n,m}| \leq \epsilon_n, 1 \leq m \leq n\}$, we have $\overline{X}_{n,m} = X_{n,m}$, and hence, $S_{n,(n\cdot)} = \overline{S}_{n,(n\cdot)}$. Thus,

$$\begin{aligned}
||S_{n,(n\cdot)} - \widetilde{S}_{n,(n\cdot)}||_\infty &= ||S_{n,(n\cdot)} - \overline{S}_{n,(n\cdot)} + \sum_{m=1}^{[n\cdot]} E\left(\overline{X}_{n,m}|\mathcal{F}_{n,m-1}\right)||_\infty \\
&\leq \sum_{m=1}^{[n\cdot]} \left|E\left(\overline{X}_{n,m}|\mathcal{F}_{n,m-1}\right)\right| \to_P 0.
\end{aligned}$$

Now, to complete the proof, we have to show that Lemma 5.3 holds on $\{|X_{n,m}| > \epsilon_n, 1 \leq m \leq n\}$. It suffices to show that

Lemma 5.5:

$$P\left(|X_{n,m}| > \epsilon_n, \text{ for some } m, 1 \leq m \leq n\right) \to 0.$$

To prove Lemma 5.5, we use Dvoretsky's proposition.

Proposition 5.1 (Dvoretsky): Let $\{\mathcal{G}_n\}$ be a sequence of σ-fields with $\mathcal{G}_n \subset \mathcal{G}_{n+1}$. If $A_n \in \mathcal{G}_n$ for each n, then for each $\delta \geq 0$, measurable with respect to $\mathcal{G}_0$,

$$P\left(\bigcup_{m=1}^{n} A_m|\mathcal{G}_0\right) \leq \delta + P\left(\sum_{m=1}^{n} P(A_m|\mathcal{G}_{m-1}) > \delta|\mathcal{G}_0\right) \tag{5.4}$$

Proof: We use induction.
(i) $n = 1$
We want to show

$$P\Big(A_1|\mathcal{G}_0\Big) \leq \delta + P\Big(P(A_1|\mathcal{G}_0) > \delta|\mathcal{G}_0\Big). \tag{5.5}$$

Consider $\Omega_\ominus = \{\omega : P(A_1|\mathcal{G}_0) \leq \delta\}$. Then (5.5) holds. Also, consider $\Omega_\oplus = \{\omega : P(A_1|\mathcal{G}_0) > \delta\}$. Then

$$\begin{aligned} P\Big(P(A_1|\mathcal{G}_0) > \delta|\mathcal{G}_0\Big) &= E\Big(\mathbf{1}\Big(P(A_1|\mathcal{G}_0) > \delta\Big)|\mathcal{G}_0\Big) \\ &= \mathbf{1}\Big(P(A_1|\mathcal{G}_0) > \delta\Big) \\ &= 1, \end{aligned}$$

and hence (5.5) also holds.
(ii) $n > 1$
Consider $\omega \in \Omega_\oplus$. Then

$$\begin{aligned} P\Big(\sum_{m=1}^{n} P(A_m|\mathcal{G}_{m-1}) > \delta|\mathcal{G}_0\Big) &\geq P\Big(P(A_1|\mathcal{G}_0) > \delta|\mathcal{G}_0\Big) \\ &= 1_{\Omega_\oplus}(\omega) \\ &= 1. \end{aligned}$$

Then, (5.4) holds. Consider $\omega \in \Omega_\ominus$. Let $B_m = A_m \cap \Omega_\ominus$. Then, for $m \geq 1$,

$$\begin{aligned} P(B_m|\mathcal{G}_{m-1}) &= P(A_m \cap \Omega_\ominus|\mathcal{G}_{m-1}) \\ &= P(A_m|\mathcal{G}_{m-1}) \cdot P(\Omega_\ominus|\mathcal{G}_{m-1}) \\ &= P(A_m|\mathcal{G}_{m-1}) \cdot 1_{\Omega_\ominus}(\omega) \\ &= P(A_m|\mathcal{G}_{m-1}). \end{aligned}$$

Now suppose $\gamma = \delta - P(B_1|\mathcal{G}_0) \geq 0$ and apply the last result for $n - 1$ sets (induction hypothesis).

$$P\Big(\bigcup_{m=2}^{n} B_m|\mathcal{G}_1\Big) \leq \gamma + P\Big(\sum_{m=2}^{n} P(B_m|\mathcal{G}_{m-1}) > \gamma|\mathcal{G}_1\Big).$$

Recall $E\Big(E(X|\mathcal{G}_0)|\mathcal{G}_1\Big) = E(X|\mathcal{G}_0)$ if $\mathcal{G}_0 \subset \mathcal{G}_1$. Taking conditional expectation w.r.t. $\mathcal{G}_0$ and noting $\gamma \in \mathcal{G}_0$,

$$P\Big(\bigcup_{m=2}^{n} B_m|\mathcal{G}_0\Big) \leq P\Big(\gamma + P\Big(\sum_{m=2}^{n} P(B_m|\mathcal{G}_{m-1}) > \gamma|\mathcal{G}_1\Big)|\mathcal{G}_0\Big)$$

$$= \gamma + P\Big(\sum_{m=2}^{n} P(B_m|\mathcal{G}_{m-1}) > \gamma|\mathcal{G}_0\Big)$$
$$= \gamma + P\Big(\sum_{m=1}^{n} P(B_m|\mathcal{G}_{m-1}) > \delta|\mathcal{G}_0\Big).$$

Since $\cup B_m = (\cup A_m) \cap \Omega_\ominus$, on $\Omega_\ominus$ we have

$$\sum_{m=1}^{n} P(B_m|\mathcal{G}_{m-1}) = \sum_{m=1}^{n} P(A_m|\mathcal{G}_{m-1}).$$

Thus, on $\Omega_\ominus$,

$$P\Big(\bigcup_{m=2}^{n} A_m|\mathcal{G}_0\Big) \le \delta - P(A_1|\mathcal{G}_0) + P\Big(\sum_{m=1}^{n} P(A_m|\mathcal{G}_{m-1}) > \delta|\mathcal{G}_0\Big).$$

The result follows from

$$P\Big(\bigcup_{m=1}^{n} A_m|\mathcal{G}_0\Big) \le P(A_1|\mathcal{G}_0) + P\Big(\bigcup_{m=2}^{n} A_m|\mathcal{G}_0\Big)$$

using monotonicity of conditional expectation and $1_{A\cup B} \le 1_A + 1_B$.

Proof of Lemma 5.5: Let $A_m = \{|X_{n,m}| > \epsilon_n\}$, $\mathcal{G}_m = \mathcal{F}_{n,m}$, and let δ be a positive number. Then, Proposition 5.1 implies

$$P(|X_{n,m}| > \epsilon_n \text{ for some } m \le n) \le \delta + P\Big(\sum_{m=1}^{n} P(|X_{n,m}| > \epsilon_n|\mathcal{F}_{n,m-1}) > \delta\Big).$$

To estimate the right-hand side, we observe that "Chebyshev's inequality" implies

$$\sum_{m=1}^{n} P(|X_{n,m}| > \epsilon_n|\mathcal{F}_{n,m-1}) \le \epsilon_n^{-2} \sum_{m=1}^{n} E(\hat{X}_{n,m}^2|\mathcal{F}_{n,m-1}) \to 0$$

so $\limsup P(|X_{n,m}| > \epsilon_n$ for some $m \le n) \le \delta$. Since δ is arbitrary, the proof of lemma and theorem is complete.

Theorem 5.4 (Martingale cental limit theorem): Let $\{X_n, \mathcal{F}_n\}$ be a martingale difference sequence, $X_{n,m} = X_m/\sqrt{n}$, and $V_k = \sum_{n=1}^{k} E\Big(X_n^2\Big|\mathcal{F}_{n-1}\Big)$. Assume that

1. $V_k/k \to_P \sigma^2$,
2. $n^{-1} \sum_{m\le n} E\Big(X_m^2 1\Big(|X_m| > \epsilon\sqrt{n}\Big)\Big) \to 0$.

Then,

$$\frac{S_{(n\cdot)}}{\sqrt{n}} \Rightarrow \sigma W(\cdot).$$

Definition 5.2: A process $\{X_n, n \geq 0\}$ is called stationary if

$$\{X_m, X_{m+1}, \ldots, X_{m+k}\} =_{\mathscr{D}} \{X_0, \ldots, X_k\}$$

for any m, k.

Definition 5.3: Let $(\Omega, \mathcal{F}, P)$ be a probability space. A measurable map $\varphi : \Omega \to \Omega$ is said to be measure preserving if $P(\varphi^{-1}A) = P(A)$ for all $A \in \mathcal{F}$.

Theorem 5.5: If $X_0, X_1, \ldots$ is a stationary sequence and $g : \mathbf{R}^{\mathbf{N}} \to \mathbf{R}$ is measurable, then $Y_k = g(X_k, X_{k+1}, \ldots)$ is a stationary sequence.

Proof: If $\mathbf{x} \in \mathbf{R}^{\mathbf{N}}$, let $g_k(\mathbf{x}) = g(x_k, x_{k+1}, \ldots)$, and if $B \in \mathcal{R}^{\mathbf{N}}$ let

$$A = \{\mathbf{x} : (g_0(\mathbf{x}), g_1(\mathbf{x}), \ldots) \in B\}.$$

To check stationarity now, we observe:

$$\begin{aligned}
P\Big(\{\omega : (Y_0, Y_1, \ldots) \in B\}\Big) &= P\Big(\{\omega : (g(X_0, X_1, \ldots), g(X_1, X_2, \ldots), \ldots) \in B\}\Big) \\
&= P\Big(\{\omega : (g_0(\mathbf{X}), g_1(\mathbf{X}), \ldots) \in B\}\Big) \\
&= P\Big(\{\omega : (X_0, X_1, \ldots) \in A\}\Big) \\
&= P\Big(\{\omega : (X_k, X_{k+1}, \ldots) \in A\}\Big) \\
&\qquad (\text{ since } X_0, X_1, \ldots \text{ is a stationary sequence}) \\
&= P\Big(\{\omega : (Y_k, Y_{k+1}, \ldots) \in B\}\Big).
\end{aligned}$$

Exercise: Show the above equality.

Definition 5.4: Assume that θ is measure preserving. A set $A \in \mathcal{F}$ is said to be invariant if $\theta^{-1}A = A$. Denote by $\mathcal{I} = \{A \text{ in } \mathcal{F}, A = \theta^{-1}A\}$ in measure.

Definition 5.5: A measure preserving transformation on $(\Omega, \mathcal{F}, P)$ is said to be ergodic if $\mathcal{I}$ is trivial, i.e., for every $A \in \mathcal{I}$, $P(A) \in \{0, 1\}$.

Example: Let $X_0, X_1, \ldots$ be the i.i.d. sequence. If $\Omega = \mathbf{R}^{\mathbf{N}}$ and θ is the shift operator, then an invariant set A has $\{\omega : \omega \in A\} = \{\omega : \theta\omega \in A\} \in \sigma(X_1, X_2, \ldots)$. Iterating gives

$$A \in \bigcap_{n=1}^{\infty} \sigma(X_n, X_{n+1}, \ldots) = \mathcal{T}, \text{ the tail } \sigma\text{-field}$$

so $\mathcal{I} \subset \mathcal{T}$. For an i.i.d. sequence, Kolmogorov's 0-1 law implies $\mathcal{T}$ is trivial, so $\mathcal{I}$ is trivial and the sequence is ergodic. We call θ the ergodic transformation.

Theorem 5.6: Let $g : \mathbf{R}^{\mathbf{N}} \to \mathbf{R}$ be measurable. If $X_0, X_1, \ldots$ is an ergodic stationary sequence, then $Y_k = g(X_k, X_{k+1}, \ldots)$ is ergodic.

Proof: Suppose $X_0, X_1, \ldots$ is defined on sequence space with $X_n(\omega) = \omega_n$. If B has $\{\omega : (Y_0, Y_1, \ldots) \in B\} = \{\omega : (Y_1, Y_2, \ldots) \in B\}$, then $A = \{\omega : (Y_0, Y_1, \ldots) \in B\}$ is shift invariant.

Theorem 5.7 (Birkhoff's ergodic theorem): For any $f \in L_1(P)$,

$$\frac{1}{n}\sum_{m=0}^{n-1} f(\theta^m \omega) \to E(f|\mathcal{G}) \text{ a.s. and in } L_1(P),$$

where θ is measure preserving transformation on $(\Omega, \mathcal{F}, P)$ and $\mathcal{G} = \{A \in \mathcal{F} : \theta^{-1} A = A\}$.

For the proof, see [8].

Theorem 5.8: Suppose $\{X_n, n \in \mathbf{Z}\}$ is an ergodic stationary sequence of martingale differences, i.e., $\sigma^2 = EX_n^2 < \infty$ and $E(X_n|\mathcal{F}_{n-1}) = 0$ with respect to $\mathcal{F}_n = \sigma(X_m, m \le n)$. Let $S_n = X_1 + \cdots + X_n$. Then,

$$\frac{S_{(n\cdot)}}{\sqrt{n}} \Rightarrow \sigma W(\cdot).$$

Proof: Let $u_n = E(X_n^2|\mathcal{F}_{n-1})$. Then u_n can be written as $\theta(X_{n-1}, X_{n-2}, \ldots)$, and hence, by Theorem 5.6 u_n is stationary and ergodic. By Birkhoff's ergodic theorem ($\mathcal{G} = \{\emptyset, \Omega\}$),

$$\frac{1}{n}\sum_{m=1}^{n} u_m \to Eu_0 = EX_0^2 = \sigma^2 \text{ a.s.}$$

The last conclusion shows that (i) of Theorem 5.4 holds. To show (ii), we observe

$$\begin{aligned}\frac{1}{n}\sum_{m=1}^{n} E\left(X_m^2 1\left(|X_m| > \epsilon\sqrt{n}\right)\right) &= \frac{1}{n}\sum_{m=1}^{n} E\left(X_0^2 1\left(|X_0| > \epsilon\sqrt{n}\right)\right) \\ &\qquad \text{(because of stationarity)} \\ &= E\left(X_0^2 1\left(|X_0| > \epsilon\sqrt{n}\right)\right) \to 0\end{aligned}$$

by the dominated convergence theorem. This completes the proof.

Let's consider stationary process, (with $\xi_i i.i.d.$, $E\xi_i = 0$ and $E\xi_i^2 < \infty$)

$$X_n = \sum_{k=0}^{\infty} c_k \xi_{n-k}$$

with

$$\sum_{k=0}^{\infty} c_k^2 < \infty.$$

If ξ_i are i.i.d., X_n is definitely stationary, but it is not martingale difference process. This is called moving average process. What we will do is we start with stationary ergodic process, and then we will show that limit of this process is the limit of martingale difference sequence. Then this satisfies the conditions of martingale central limit theorem. Note that in our example, $EX_n^2 < \infty$ for all n.

If X_n is stationary second-order, then $r(n) = EX_n\overline{X}_o$ is positive definite.

We can separate phenomenon into two parts: new information(non-deterministic) and non-new information (deterministic).

$$EX_nX_0 = \int_0^{2\pi} e^{in\lambda} dF(\lambda),$$

where F is the spectral measure. In case $X_n = \sum_{k=0}^{\infty} c_k\xi_{n-k}$, then $F \ll$ Lebesgue measure.

In fact, we have the following theorem:

Theorem 5.9: There exist $\bar{c}_k$ and ϕ such that
- $f(e^{i\lambda}) = \left|\phi(e^{i\lambda})\right|^2$
- $\phi(e^{i\lambda}) = \sum_{k=0}^{\infty} \bar{c}_k e^{ik\lambda}$

if and only if

$$\int_0^{2\pi} \log f(\lambda) dF(\lambda) > -\infty.$$

Now we start with $\{X_n : n \in \mathbf{Z}\}$ ergodic stationary sequence such that
- $EX_n = 0, EX_n^2 < \infty.$
- $\sum_{n=1}^{\infty} ||E(X_0|\mathcal{F}_{-n})||_2 < \infty.$

The idea is if we go back, then X_n will be independent of X_0.

Let

$$H_n = \{Y \in \mathcal{F}_n \text{ with } EY^2 < \infty\} = L^2(\Omega, \mathcal{F}_n, P)$$
$$K_n = \{Y \in H_n \text{ with } EYZ = 0 \text{ for all } Z \in H_{n-1}\} = H_n \ominus H_{n-1}.$$

Geometrically, $H_0 \supset H_{-1} \supset H_{-2} \dots$ is a sequence of subspaces of L^2 and K_n is the orthogonal complement of H_{n-1}. If Y is a random variable, let

$$(\theta^n Y)(\omega) = Y(\theta^n \omega),$$

i.e., θ is isometry(measure-preserving) on L^2. Generalizing from the example $Y = f(X_{-j}, \ldots, X_k)$, which has $\theta^n Y = f(X_{n-j}, \ldots, X_{n+k})$, it is easy to see that if $Y \in H_k$, then $\theta^n Y \in H_{k+n}$, and hence $Y \in K_j$ then $\theta^n Y \in K_{n+j}$.

Lemma 5.6: Let P be a projection such that $X_j \in H_{-j}$ implies $P_{H_{-j}} X_j = X_j$. Then,

$$\begin{aligned} \theta P_{H_{-1}} X_j &= P_{H_0} X_{j+1} \\ &= P_{H_0} \theta X_j. \end{aligned}$$

Proof: For $j \le -1$,

$$\theta \underbrace{P_{H_{-j}} X_j}_{X_j} = \theta X_j = X_{j+1}.$$

We will use this property. For $Y \in H_{-1}$,

$$\begin{aligned} & X_j - P_{H_{-1}} X_j \perp Y \\ \Rightarrow\ & (X_j - P_{H_{-1}} X_j, Y)_2 = 0 \\ \Rightarrow\ & \left(\theta(X_j - P_{H_{-1}} X_j), \theta Y\right)_2 = 0 \text{ (since } \theta \text{ is isometry on } L^2) \end{aligned}$$

Since $Y \in H_{-1}$, θY generates H_0. Therefore, for all $Z \in H_0$, we have

$$\begin{aligned} & \left((\theta X_j - \theta P_{H_{-1}} X_j), Z\right)_2 = 0 \\ \Rightarrow\ & (\theta X_j - \theta P_{H_{-1}} X_j) \perp Z \\ \Rightarrow\ & \theta P_{H_{-1}} X_j = P_{H_0} \theta X_j = P_{H_0} X_{j+1}. \end{aligned}$$

We come now to the central limit theorem for stationary sequences.

Theorem 5.10: Suppose $\{X_n, n \in \mathbf{Z}\}$is an ergodic stationary sequence with $EX_n = 0$ and $EX_n^2 < \infty$. Assume

$$\sum_{n=1}^{\infty} ||E(X_0|\mathcal{F}_{-n})||_2 < \infty$$

Let $S_n = X_1 + \ldots + X_n$. Then

$$\frac{S_{(n\cdot)}}{\sqrt{n}} \Rightarrow \sigma W(\cdot)$$

where we do not know what σ is.

Proof: If X_0 happened to be in K_0 since $X_n = \theta^n X_0 \in K_n$ for all n, and taking $Z = 1_A \in H_{n-1}$, we would have $E(X_n 1_A) = 0$ for all $A \in \mathcal{F}_{n-1}$ and hence $E(X_n|\mathcal{F}_{n-1}) = 0$. The next best thing to having $X_n \in K_0$ is to have

$$X_0 = Y_0 + Z_0 - \theta Z_0 \quad (*)$$

with $Y_0 \in K_0$ and $Z_0 \in L^2$. Let

$$\begin{aligned}
Z_0 &= \sum_{j=0}^{\infty} E(X_j|\mathcal{F}_{-1}) \\
\theta Z_0 &= \sum_{j=0}^{\infty} E(X_{j+1}|\mathcal{F}_0) \\
Y_0 &= \sum_{j=0}^{\infty} \Big(E(X_j|\mathcal{F}_0) - E(X_j|\mathcal{F}_{-1}) \Big).
\end{aligned}$$

Then we can solve $(*)$ formally

$$Y_0 + Z_0 - \theta Z_0 = E(X_0|\mathcal{F}_0) = X_0. \tag{5.6}$$

We let

$$S_n = \sum_{m=1}^{n} X_m = \sum_{m=1}^{n} \theta^m X_0 \quad \text{and } T_n = \sum_{m=1}^{n} \theta^m Y_0.$$

We want to show that T_n is martingale difference sequence. We have $S_n = T_n + \theta Z_0 - \theta^{n+1} Z_0$. The $\theta^m Y_0$ are a stationary ergodic martingale difference sequence (ergodicity follows from Theorem 5.6), so Theorem 5.8 implies

$$\frac{T_{(n\cdot)}}{\sqrt{n}} \Rightarrow \sigma W(\cdot), \text{ where } \sigma^2 = EY_0^2.$$

To get rid of the other term, we observe

$$\frac{\theta Z_0}{\sqrt{n}} \to 0 \text{ a.s.}$$

and

$$\begin{aligned}
P\Big(\max_{0 \le m \le n-1} |\theta^{m+1} Z_0| > \epsilon\sqrt{n}\Big) &\le \sum_{m=0}^{n-1} P\Big(|\theta^{m+1} Z_0| > \epsilon\sqrt{n}\Big) \\
&= nP\Big(|Z_0| > \epsilon\sqrt{n}\Big) \\
&\le \epsilon^{-2} E\Big(Z_0^2 \mathbf{1}_{\{|Z_0| > \epsilon\sqrt{n}\}}\Big) \to 0.
\end{aligned}$$

The last inequality follows from the stronger form of Chevyshev,

$$E\left(Z_0^2 \mathbf{1}_{\{|Z_0|>\epsilon\sqrt{n}\}}\right) \geq \epsilon^2 n P\left(|Z_0| > \epsilon\sqrt{n}\right).$$

Therefore, in view of the above comments,

$$\begin{aligned} & \frac{S_n}{\sqrt{n}} = \frac{T_n}{\sqrt{n}} + \frac{\theta Z_0}{\sqrt{n}} - \frac{\theta^{n+1} Z_0}{\sqrt{n}} \\ \Rightarrow\ & \frac{S_n}{\sqrt{n}} - \frac{T_n}{\sqrt{n}} \xrightarrow{p} 0 \\ \Rightarrow\ & \lim_{n\to\infty} \frac{S_{(n\cdot)}}{\sqrt{n}} = \lim_{n\to\infty} \frac{T_{(n\cdot)}}{\sqrt{n}} = \sigma W(\cdot). \end{aligned}$$

Theorem 5.11: Suppose $\{X_n, n \in \mathbf{Z}\}$ is an ergodic stationary sequence with $EX_n = 0$ and $EX_n^2 < \infty$. Assume

$$\sum_{n=1}^{\infty} ||E(X_0|\mathcal{F}_{-n})||_2 < \infty.$$

Let $S_n = X_1 + \ldots + X_n$. Then

$$\frac{S_{(n\cdot)}}{\sqrt{n}} \Rightarrow \sigma W(\cdot),$$

where

$$\sigma^2 = EX_0^2 + 2\sum_{n=1}^{\infty} EX_0 X_n.$$

If $\sum_{n=1}^{\infty} EX_0X_n$ diverges, theorem will not be true. We will show that $\sum_{n=1}^{\infty} |EX_0X_n| < \infty$. This theorem is different from previous theorem since we now specify σ^2.

Proof: First,

$$\begin{aligned} \left|EX_0X_m\right| &= \left|E\left(E(X_0X_m|\mathcal{F}_0)\right)\right| \\ &\leq E\left|X_0E(X_m|\mathcal{F}_0)\right| \\ &\leq ||X_0||_2 \cdot ||E(X_m|\mathcal{F}_0)||_2 \text{ (by Cauchy Schwarz inequality)} \\ &= ||X_0||_2 \cdot ||E(X_0|\mathcal{F}_{-m})||_2 \text{ (by shift invariance).} \end{aligned}$$

Therefore, by assumption,

$$\sum_{n=1}^{\infty} |EX_0X_n| \leq ||X_0||_2 \sum_{n=1}^{\infty} ||E(X_0|\mathcal{F}_{-m})||_2 < \infty.$$

Next,

$$ES_n^2 = \sum_{j=1}^{n}\sum_{k=1}^{n} EX_jX_k$$
$$= nEX_0^2 + 2\sum_{m=1}^{n-1}(n-m)EX_0X_m.$$

From this, it follows easily that

$$\frac{ES_n^2}{n} \to EX_0^2 + 2\sum_{m=1}^{\infty} EX_0X_m.$$

To finish the proof, let $T_n = \sum_{m=1}^{n} \theta^m Y_0$, observe $\sigma^2 = EY_0^2$, and

$$n^{-1}E(S_n - T_n)^2 = n^{-1}E(\theta Z_0 - \theta^{n+1}Z_0)^2$$
$$\leq \frac{3EZ_0^2}{n} \to 0$$

since $(a-b)^2 \leq (2a)^2 + (2b)^2$.

We proved central limit theorem of ergodic stationary process. We will discuss examples: M-dependence and moving average.

Example 1. M-dependent sequences: Let X_n, $n \in \mathbb{Z}$ be a stationary sequence with $EX_n = 0, EX_n^2 < \infty$. Assume that $\sigma(\{X_j, j \leq 0\})$ and $\sigma(\{X_j, j \geq M\})$ are independent. In this case, $E(X_0|\mathcal{F}_{-n}) = 0$ for $n > M$, and $\sum_{n=0}^{\infty} ||E(X_0|\mathcal{F}_{-n})||_2 < \infty$. Let $\mathcal{F}_{-\infty} = \cap_m \sigma(\{X_j, j \geq M\})$ and $\mathcal{F}_k = \sigma(\{X_j, j \leq k\})$. If $m - k > M$, then $\mathcal{F}_{-\infty} \perp \mathcal{F}_k$. Recall Kolmogorov 0-1 law. If $A \in \mathcal{F}_k$ and $B \in \mathcal{F}_{-\infty}$, then $P(A \cap B) = P(A) \cdot P(B)$. For all $A \in \cup_k \mathcal{F}_k$, $A \in \sigma(\cup_k \mathcal{F}_k)$. Also, $A \in \mathcal{F}_{-\infty}$, where $\mathcal{F}_{-\infty} \subset \cup_k \mathcal{F}_{-k}$. Therefore, by Kolmogorov 0-1 law, $P(A \cap A) = P(A) \cdot P(A)$, and hence, $\{X_n\}$ is stationary and ergodic. Thus, Theorem 5.8 implies

$$\frac{S_{n,(n\cdot)}}{\sqrt{n}} \Rightarrow \sigma W(\cdot),$$

where

$$\sigma^2 = E_0^2 + 2\sum_{m=1}^{M} EX_0X_m.$$

Example 2. Moving average: Suppose

$$X_m = \sum_{k\geq 0} c_k \xi_{m-k}, \text{ where } \sum_{k\geq 0} c_k^2 < \infty,$$

and ξ_i, $i \in \mathbf{Z}$ are i.i.d. with $E\xi_i = 0$ and $E\xi_i^2 = 1$. Clearly, $\{X_n\}$ is a stationary sequence since series converges. Check whether $\{X_n\}$ is ergodic. We have

$$\bigcap_n \sigma(\{X_m, m \le n\}) \subset \bigcap_n \sigma(\{\xi_k, k \le n\});$$

therefore, by Kolmogorov 0-1 law, $\{X_n\}$ is ergodic. Next, if $\mathcal{F}_{-n} = \sigma(\{\xi_m, m \le -n\})$, then

$$\begin{aligned} ||E(X_0|\mathcal{F}_{-n})||_2 &= ||\sum_{k \ge n} c_k \xi_k||_2 \\ &= \Big(\sum_{k \ge n} c_k^2\Big)^{1/2}. \end{aligned}$$

If, for example, $c_k = (1+k)^{-p}$, $||E(X_0|\mathcal{F}_{-n})||_2 \sim n^{(1/2-p)}$, and Theorem 5.11 applies if $p > 3/2$.

Let $\mathcal{G}, \mathcal{H} \subset \mathcal{F}$, and

$$\alpha(\mathcal{G}, \mathcal{H}) = \sup_{A \in \mathcal{G}, B \in \mathcal{H}} \Big\{|P(A \cap B) - P(A)P(B)|\Big\}.$$

If $\alpha = 0$, $\mathcal{G}$, and $\mathcal{H}$ are independent, α measures the dependence of two σ-algebras.

Lemma 5.7: Let $p, q, r \in (1, \infty]$ with $1/p + 1/q + 1/r = 1$, and suppose $X \in \mathcal{G}$, $Y \in \mathcal{H}$ have $E|X|^p, E|Y|^q < \infty$. Then

$$|EXY - EXEY| \le 8||X||_p||Y||_q\Big(\alpha(\mathcal{G}, \mathcal{H})\Big)^{1/r}.$$

Here, we interpret $x^0 = 1$ for $x > 0$ and $0^0 = 0$.

Proof: If $\alpha = 0$, X and Y are independent and the result is true, so we can suppose $\alpha > 0$. We build up to the result in three steps, starting with the case $r = \infty$.
(a). $r = \infty$

$$|EXY - EXEY| \le 2||X||_p||Y||_q.$$

Proof of (a): Hölder's inequality implies $|EXY| \le ||X||_p||Y||_q$, and Jensen's inequality implies

$$||X||_p||Y||_q \ge \Big|E|X|E|Y|\Big| \ge |EXEY|;$$

thus, the result follows from the triangle inequality.

(b) $X, Y \in L^\infty$

$$|EXY - EXEY| \le 4||X||_\infty ||Y||_\infty \alpha(\mathcal{G}, \mathcal{H}).$$

Proof of (b): Let $\eta = sgn\Big(E(Y|\mathcal{G}) - EY\Big) \in \mathcal{G}$. $EXY = E(XE(Y|\mathcal{G}))$, so

$$\begin{aligned} |EXY - EXEY| &= |E(X(E(Y|\mathcal{G}) - EY))| \\ &\le ||X||_\infty E|E(Y|\mathcal{G}) - EY| \\ &= ||X||_\infty E(\eta E(Y|\mathcal{G}) - EY) \\ &= ||X||_\infty (E(\eta Y) - E\eta EY). \end{aligned}$$

Applying the last result with $X = Y$ and $Y = \eta$ gives

$$|E(Y\eta) - EYE\eta| \le ||Y||_\infty |E(\zeta\eta) - E\zeta E\eta|,$$

where $\zeta = sgn(E(\eta|\mathcal{H}) - E\eta)$. Now $\eta = 1_A - 1_B$ and $\zeta = 1_C - 1_D$, so

$$\begin{aligned} |E(\zeta\eta) - E\zeta E\eta| &= |P(A \cap C) - P(B \cap C) - P(A \cap D) + P(B \cap D) \\ &\quad -P(A)P(C) + P(B)P(C) + P(A)P(D) - P(B)P(D)| \\ &\le 4\alpha(\mathcal{G}, \mathcal{H}). \end{aligned}$$

Combining the last three displays gives the desired result.
(c) $q = \infty$, $1/p + 1/r = 1$

$$|EXY - EXEY| \le 6||X||_p ||Y||_\infty \Big(\alpha(\mathcal{G}, \mathcal{H})\Big)^{1-1/p}.$$

Proof of (c): Let $C = \alpha^{-1/p}||X||_p$, $X_1 = X1_{(|X| \le C)}$, and $X_2 = X - X_1$.

$$\begin{aligned} |EXY - EXEY| &\le |EX_1Y - EX_1EY| + |EX_2Y - EX_2EY| \\ &\le 4\alpha C||Y||_\infty + 2||Y||_\infty E|X_2| \end{aligned}$$

by (a) and (b). Now

$$E|X_2| \le C^{-(p-1)}E(|X|^p 1_{(|X| \le C)}) \le C^{-p+1}E|X|^p.$$

Combining the last two inequalities and using the definition of C gives

$$|EXY - EXEY| \le 4\alpha^{1-1/p}||X||_p ||Y||_\infty + 2||Y||_\infty \alpha^{1-1/p}||X||_p^{-p+1+p},$$

which is the desired result.

Finally, to prove Lemma 5.7, let $C = \alpha^{-1/q}||Y||_q$, $Y_1 = Y1_{(|Y| \le C)}$, and $Y_2 = Y - Y_1$.

$$\begin{aligned} |EXY - EXEY| &\le |EXY_1 - EXEY_1| + |EXY_2 - EXEY_2| \\ &\le 6C||X||_p \alpha^{1-1/p} + 2||X||_p ||Y_2||_\theta, \end{aligned}$$

where $\theta = (1 - 1/p)^{-1}$ by (c) and (a). Now

$$E|Y|^\theta \le C^{-q+\theta} E(|Y|^q 1_{(|Y| \le C)}) \le C^{-1+\theta} E|Y|^q.$$

Taking the $1/\theta$ root of each side and recalling the definition of C

$$||Y_2||_\theta \le C^{-(q-\theta)} ||Y||_q^{q/\theta} \le \alpha^{(q-\theta)/q\theta} ||Y||_q,$$

so we have

$$|EXY - EXEY| \le 6\alpha^{-1/q} ||Y||_q ||X||_p \alpha^{1-1/p} + 2||X||_p \alpha^{1/\theta - 1/q} ||Y||_q^{1/\theta + 1/q},$$

proving Lemma 5.7.

Combining Theorem 5.11 and Lemma 5.7 gives:

Theorem 5.12: Suppose X_n, $n \in \mathbf{Z}$ is an ergodic stationary sequence with $EX_n = 0$, $E|X_0|^{2+\delta} < \infty$. Let $\alpha(n) = \alpha(\mathcal{F}_{-n}, \sigma(X_0))$, where $\mathcal{F}_{-n} = \sigma(\{X_m, m \le -n\})$, and suppose

$$\sum_{n=1}^{\infty} \alpha(n)^{\delta/2(2+\delta)} < \infty.$$

If $S_n = X_1 + \cdots X_n$, then

$$\frac{S_{(n\cdot)}}{\sqrt{n}} \Rightarrow \sigma W(\cdot),$$

where

$$\sigma^2 = EX_0^2 + 2\sum_{n=1}^{\infty} EX_0 X_n.$$

Proof: To use Lemma 5.7 to estimate the quantity in Theorem 5.11 we begin with

$$||E(X|\mathcal{F})||_2 = \sup\{E(XY) : Y \in \mathcal{F}, ||Y||_2 = 1\} \quad (*).$$

Proof of (*): If $Y \in \mathcal{F}$ with $||Y||_2 = 1$, then using a by now familiar property of conditional expectation and the Cauchy-Schwarz inequality

$$EXY = E(E(XY|\mathcal{F})) = E(YE(X|\mathcal{F})) \le ||E(X|\mathcal{F})||_2 ||Y||_2$$

Equality holds when $Y = E(X|\mathcal{F})/||E(X|\mathcal{F})||_2$.

Letting $p = 2 + \delta$ and $q = 2$ in Lemma 5.7, noticing

$$\frac{1}{r} = 1 - \frac{1}{p} - \frac{1}{q} = \frac{\delta}{2(2+\delta)}$$

and recalling $EX_0 = 0$, showing that if $Y \in \mathcal{F}_{-n}$

$$|EX_0 Y| \le 8\|X_0\|_{2+\delta}\|Y\|_2 \alpha(n)^{\delta/2(2+\delta)}.$$

Combining this with (*) gives

$$\|E(X_0|\mathcal{F}_{-n})\|_2 \le 8\|X_0\|_{2+\delta}\alpha(n)^{\delta/2(2+\delta)},$$

and it follows that the hypotheses of Theorem 5.11 are satisfied.

6 Empirical process

We have so far considered the weak convergence of stochastic processes with values in complete separable metric spaces. The main assumption needed to study such convergence in terms of bounded continuous functions is that the measures be defined on Borel subsets of the space. This influences theorems like the Prokhorov theorem. In other words, the random variables with values in separable metric spaces. If we consider the space $D[0,1]$ with sup norm, we get non-separability. We note that if we consider the case of empirical processes

$$X_n(t) = \sqrt{n}(\hat{F}_n(t) - t),$$

where $\hat{F}_n(t)$ is the empirical distribution function of i.i.d. uniform random variable, then the above X_n is not a Borel measureable function in $D[0,1]$ with sup norm if we consider the domain of $X_n(.,w)$ with $w \epsilon [0,1]$? Thus, the weak convergence definition above cannot be used in this case. Dudley [6] developed an alternative weak convergence theory. Even this cannot handle the general empirical processes, hence some statistical applications. One therefore has to introduce outer expectations of X_n of possibly non-measureable maps as far as limit of X_n is Borel measureable. We shall also consider convergence in probability and almost sure convergence for such functions. We therefore follow this idea as in [25]. A similar idea was exploited in the basic paper on invariance principle in non-separable Banach spaces by Dudley and Phillips ([7], see also [18]).

Let $(\Omega, \mathcal{A}, P)$ be an arbitrary probability space and $T : \Omega \to \bar{\mathbf{R}}$ an arbitrary map. The *outer integral* of T with respect to P is defined as

$$E^* T = \inf\{EU : U \geq T, U : \Omega \to \bar{\mathbf{R}} \text{ measurable and } EU \text{ exists}\}.$$

Here, as usual, EU is understood to exist if at least one of EU^+ or EU^- is finite. The *outer probability* of an arbitrary subset B of Ω is

$$P^*(B) = \inf\{P(A) : A \supset B, A \in \mathcal{A}\}.$$

Note that the functions U in the definition of outer integral are allowed to take the value ∞, so that the infimum exists.

The inner integral and inner probability can be defined in a similar fashion. Equivalently, they can be defined by $E_* T = -E^*(-T)$ and $P_*(B) = 1 - P^*(B)$, respectively.

In this section, $\mathbf{D}$ is metric space with a metric d. The set of all continuous, bounded functions $f : \mathbf{D} \to \mathbf{R}$ is denoted by $C_b(\mathbf{D})$.

Definition 6.1: Let $(\Omega_\alpha, \mathcal{A}_\alpha, P_\alpha)$, $\alpha \in I$ be a net of probability spaces, $X_\alpha : \Omega_\alpha \to \mathbf{D}$, and $P_\alpha = P \circ X_\alpha^{-1}$. Then we say that the net X_α converges weakly to a Borel measure L, i.e., $P_\alpha \Rightarrow L$ if

$$E^* f(X_\alpha) \to \int f dL, \qquad \text{for every } f \in C_b(\mathbf{D}).$$

Theorem 6.1 (Portmanteau): The following statements are equivalent:

(i) $P_\alpha \Rightarrow L$;

(ii) $\liminf P_*(X_\alpha \in G) \geq L(G)$ for every open G;

(iii) $\limsup P^*(X_\alpha \in F) \leq L(F)$ for every closed set F;

(iv) $\liminf E_* f(X_\alpha) \geq \int f dL$ for every lower semicontinuous f that is bounded below;

(v) $\limsup E^* f(X_\alpha) \leq \int f dL$ for every upper semicontinuous f that is bounded above;

(vi) $\lim P^*(X_\alpha \in B) = \lim P_*(X_\alpha \in B) = L(B)$ for every Borel set B with $L(\partial B) = 0$.

(vii) $\liminf E_* f(X_\alpha) \geq \int f dL$ for every bounded, Lipschitz continuous, non-negative f.

Proof: (ii) and (iii) are equivalent by taking complements. Similarly, (ii) and (v) are equivalent by using f by $-f$. The (i) $\Rightarrow$ (vii) is trivial.

(vii) $\Rightarrow$ (ii) Suppose (vii) holds. For every open G, consider a sequence of Lipchitz continuous functions $f_m \geq 0$ and $f_m \uparrow 1_G$ ($f_m(x) = md(x, D - G) > 1$, then

$$\liminf P_*(X_\alpha \in G) \geq \liminf E_*(f(X_\alpha)) \geq \int f_m dL.$$

Letting $m \to \infty$, we get the result.

(ii) $\Rightarrow$ (iv) Let f be lower semicontinuous with $f \geq 0$. Define

$$f_m = \sum_{i=1}^{m^2} (1/m) 1_{G_i}, \; G_i = \{x : f(x) > i/m\}.$$

Then $f_m \leq f$ and $|f_m(x) - f(x)| \leq 1/m$ when $x \leq m$. Fix m, use the fact G_i is open for all i,

$$\liminf E_* f(X_\alpha) \geq \liminf E_* f_m(X_\alpha) \geq \sum_{i=1}^{m^2} \frac{1}{m} P(X \in G_i) = \int f_m dL.$$

Let $n \to \infty$, we get (iv) for non-negative lower semi continuous f. The conclusion follows.

Since a continuous function is both upper and lower continuous (iv) and (v) imply (i). We now prove (vi) is equivalent to others. Consider (ii) $\Rightarrow$ (vi). If (ii) and (iii) hold, then

$$L(int\, B) \leq \liminf P_*(X_\alpha \in int\, B) \leq \limsup P^*(X_\alpha \in B) \leq L(\overline{B}).$$

If $L(\delta B) = 0$, we get equalities giving (vi).

(vi) ⇒ (iii) Suppose (vi) holds and let F be closed. Write $F^{\epsilon} = \{x : d(X, F) < \epsilon\}$. The sets ∂F^{ϵ} are disjoint for different $\epsilon > 0$, giving at most countably many have non-zero L-measure. Choose $\epsilon_m \downarrow 0$ with $L(\partial F^{\epsilon_m}) = 0$. For fixed m,

$$\limsup P^*(X_\alpha \in F) \le \limsup P^*(X_\alpha \in \overline{F}^{\epsilon_m}) = L(\overline{F^{\epsilon_m}}).$$

Let $m \to \infty$ to complete the proof.

Definition 6.2: The net of maps X_α is asymptotically measurable if and only if

$$E^* f(X_\alpha) - E_* f(X_\alpha) \to 0, \qquad \text{for every } f \in C_b(\mathbf{D}).$$

The net X_α is *asymptotically tight* if for every $\epsilon > 0$ there exists a compact set K, such that

$$\liminf P_*(X_\alpha \in K^\delta) \ge 1 - \epsilon, \qquad \text{for every } \delta > 0.$$

Here $K^\delta = \{y \in \mathbf{D} : d(y, K) < \delta\}$ is the "δ-enlargement" around K. A collection of Borel measurable maps X_α is called uniformly tight if, for every $\epsilon > 0$, there is a compact K with $P(X_\alpha \in K) \ge 1 - \epsilon$ for every α.

The δ in the definition of tightness may seem a bit overdone. Its non-asymptotic tightness as defined is essentially weaker than the same condition but with K instead of K^δ. This is caused by a second difference with the classical concept of uniform tightness: the enlarged compacts need to contain mass $1 - \epsilon$ only in the limit.

Meanwhile, nothing is gained in simple cases: for Borel measurable maps in a Polish space, asymptotic tightness and uniform tightness are the same. It may also be noted that, although K^δ is dependent on the metric, the property of asymptotic tightness depends on the topology only. One nice consequence of the present tightness concept is that weak convergence usually implies asymptotic measurability and tightness.

Lemma 6.1: (i) $X_\alpha \Rightarrow X$, then X_α is asymptotically measurable.
(ii) If $X_\alpha \Rightarrow X$, then X_α is asymptotically tight if and only if X is tight.

Proof: (i) This follows upon applying the definition of weak convergence to both f and $-f$.

(ii) Fix $\epsilon > 0$. If X is tight, then there is a compact K with $P(X \in K) > 1 - \epsilon$. By the Portmanteau theorem, $\liminf P_*(X_\alpha \in K^\delta) \ge P(X \in K^\delta)$, which is larger than $1 - \epsilon$ for every $\delta > 0$. Conversely, if X_α is tight, then there is a compact K with $\liminf P_*(X_\alpha \in K^\delta) \ge 1 - \epsilon$. By the Portmanteau theorem, $P(X \in \overline{K^\delta}) \ge 1 - \epsilon$. Let $\delta \downarrow 0$.

The next version of Prohorov's theorem may be considered a converse of the previous lemma. It comes in two parts, one for nets and one for sequences, neither one of which implies the other. The sequence case is the deepest of the two.

Theorem 6.2 (Prohorov's theorem): (i) If the net X_α is asymptotically tight and asymptotically measurable, then it has a subnet $X_{\alpha(\beta)}$ that converges in law to a tight Borel law.

(ii) If the sequence X_n is asymptotically tight and asymptotically measurable, then it has a subsequence X_{n_j} that converges weakly to a tight Borel law.

Proof: (i) Consider $(E^* f(X_\alpha))_{f \in C_b(D)}$ as a net in product space

$$\prod_{f \in C_b(D)} [-\|f\|_\infty, \|f\|_\infty],$$

which is compact in product topology. Hence, the net has convergent subnet. This implies there exist constants $L(f) \in [-\|f\|_\infty, \|f\|_\infty]$ such that

$$E^* f(X_{\alpha(\beta)}) \to L(f) \text{ for } f \in C_b(D).$$

In view of the asymptotic measurability, the numbers are the limits of corresponding $E_* f(X_\alpha)$. Now,

$$\begin{aligned} L(f_1 + f_2) &\le \lim(E^* f_1(X_{\alpha(\beta)}) + E^* f_2(X_{\alpha(\beta)})) \\ &= L(f_1) + L(f_2) \\ &= \lim(E_* f_1(X_{\alpha(\beta)}) + E_* f_2(X_{\alpha(\beta)})) \\ &\le L(f_1 + f_2). \end{aligned}$$

Thus, $L : C_b(D) \to \mathbb{R}$ is a additive, and similarly $L(\lambda f) = \lambda L(f)$ for $\alpha \in \mathbb{R}$ and L is positive. $L(f) \ge 0$ for $f \ge 0$. If $f_m \downarrow 0$, $L(f_m) \downarrow 0$. To see this, fix $\epsilon > 0$. There is a compact set K such that $\liminf P_*(X_\alpha \in K^\delta) > 1 - \epsilon$ for all $\delta > 0$. Using the Dini theorem, for sufficiently large m, $|f_m(x)| \le \epsilon$ for all $x \in K$. Using compactness of K, there exists $\delta > 0$, such that $|f_m(x)| \le 2\epsilon$ for every $x \in K^\delta$. One gets $A_\alpha \subseteq \{X_\alpha \in K^\delta\}$ measurable $(1_{\{X_\alpha \in K^\delta\}})_* = 1_{A_\alpha}$. Hence,

$$\begin{aligned} L(f_m) &= lim E f_m(X_\alpha)^*((1_{\{X_\alpha \in K^\delta\}})_* \\ &+(1_{\{X_\alpha \notin K_\delta\}})^* \le 2\epsilon + \|f_1\|_\infty \epsilon. \end{aligned}$$

Thus, L is a measure.

(ii) For $m \in \mathbb{N}$, K_m is a compact set with $\liminf P_*(X_n \in K_m^\delta) \ge 1 - \frac{1}{m}$ for $\delta > 0$. Since K_m is compact, the space $C_b(K_m)$ and $\{f : f \in C_b(K_m), \|f(x)\| \le 1, \text{for } x \in K_m\}$ are separable. Using Tietze's theorem, every f in the unit ball of $C_b(K_m)$ can be extended to an f in the unit ball of $C_b(D)$. Hence, ball of $C_b(D)$ the restrictions of which to K_m are dense in the unit ball of $C_b(K_m)$. Pick such a countable set for every m, and let $\Im$ be countably many functions obtained. For a fixed, bounded f, there is a subsequence X_{n_j} such that $E^*f(X_{n_j})$ converges to some number. Using diagonalization, one obtains a subsequence such that

$$\overline{E}^* f(X_{n_j}) \to L(f) \text{ for } f \in \Im.$$

with $L(f) \in [-1, 1]$.

Let $f \in C_b(D)$ taking values in $[-1, 1]$. Fix $\epsilon > 0$, and m. There exists $f_m \in \Im$ with $|f(x) - f_m(x)| < \epsilon$ for $x \in K_m$. Then as before there exists $\delta > 0$ such that $|f(x) - f_m(x)| \leq \epsilon$ for every $x \in K_m^\delta$. Then,

$$\begin{aligned}|E^*f(X_n) - E^*f_m(X_n)| &\leq E|f(X_n) - f_m(X_n)|^*(1_{\{X_n \in K_m^\delta\}})_* \\ &+2P^*(X_n \notin K_m^\delta) \leq 2\epsilon + \frac{2}{m}\end{aligned}$$

for n sufficiently large. Then $E^*f(X_{n_j})$ has the property that, for $\eta > 0$, there is a converging subsequence of numbers that is eventually with distance η. This gives convergence $E^*f(X_{n_j})$ to a limit following as in the proof of (i) we can get the result.

Remark: Let $D_0 \subseteq D$ and X and X_α take values in D_0. Then $X_\alpha \to X$ as maps in D_0 iff $X_\alpha \to X$ as maps in D if D and D_0 are equipped with the same metric. This is easy from Portmaneteau theorem (ii) as each $G_0 \subseteq D_0$ open is of the form $G \cap D_0$ with G open in D.

6.1 Spaces of bounded functions

A vector lattice $\mathcal{F} \subset C_b(\mathbf{D})$ is a vector space that is closed under taking positive parts: if $f \in \mathcal{F}$, then $f = f \vee 0 \in \mathcal{F}$. Then automatically $f \vee g \in \mathcal{F}$ and $f \wedge g \in \mathcal{F}$ for every $f, g \in \mathcal{F}$. A set of functions on $\mathbf{D}$ separates points of $\mathbf{D}$ if, for every pair $x \neq y \in \mathbf{D}$, there is $f \in \mathcal{F}$ with $f(x) \neq f(y)$.

Lemma 6.2: Let L_1 and L_2 be finite Borel measures on $\mathbf{D}$.
(i) If $\int f dL_1 = \int f dL_2$ for every $f \in C_b(\mathbf{D})$, then $L_1 = L_2$.
Let L_1 and L_2 be tight Borel probability measures on $\mathbf{D}$.
(ii) If $\int f dL_1 = \int f dL_2$ for every f in a vector lattice $\mathcal{F} \subset C_b(\mathbf{D})$ that contains the constant functions and separates points of $\mathbf{D}$, then $L_1 = L_2$.

Proof: (i) For every open set G, there exists a sequence of continuous functions with $0 \leq f_m \uparrow 1_G$. Using the monotone convergence theorem, $L_1(G) = L_2(G)$ for every open set G. Since $L_1(D) = L_2(D)$, the class of sets $\{A : L_1(A) = L_2(A)\}$ is a σ-field $\supseteq$ open sets.

(ii) Fix $\epsilon >$, take K compact so that $L_1(K)L_2(K) \geq 1 - \epsilon$. Note that by the Stone Weirstvass theorem, $\Im$ (containing constant and separating points of K) lattice $\subseteq C_b(K)$ is uniformly dense in $C_b(k)$. Given $g \in C_b(D)$ with $0 \leq g \leq 1$, take $f \in \Im$, with $|g(x) - f(x)| \leq \epsilon$ for $x \in K$. Then

$$| \int gdL_1 - \int gdL_2| \leq | \int (f \wedge 1)^+ dL_1 - \inf(f \wedge 1)^+ dL_2 + \epsilon.$$

This equals 4ϵ as $(f \wedge 1)^+ \in \Im$. Hence, one can equals 4ϵ as $(f \wedge 1)^+ \in \Im$. Hence, one can prove $\int gdL_1 = \int gdL_2$ for all $g \in C_b(D)$.

Lemma 6.3: Let the net X_α be asymptotically tight, and suppose $E^* f(X_\alpha) - E_* f(X_\alpha) \to 0$ for every f in a subalgebra $\mathcal{F}$ of $C_b(\mathbf{D})$ that separates points of $\mathbf{D}$. Then the net X_α is asymptotically measurable.

Proof: Fix $\epsilon > 0$ and K compact such that lim sup $P^*(X_\alpha \notin K^\delta) \leq \epsilon$ for all $\delta > 0$. Assume without loss of generality that $\Im$ contains constant functions. Then restrictions of the function $\Im$ into K are uniformly dense in $C_b(K)$ as before. Hence, given $f \in C_b(D)$, there exists $g \in \Im$ with $|f(x) - g(x)| < \frac{\epsilon}{4}$ for $x \in K$. As K is compact one can show there exists a $\delta > 0$ such that $|f(x) - g(x)| < \frac{\epsilon}{3}$ for $x \in K^\delta$. Let $\{X_\alpha \in k^\delta\}_*$ be a measurable set. Then $P(\Omega_\alpha \setminus \{X_\alpha \in K^\delta\}_*) = P^*\{X_\alpha \notin K^\delta)$ and for α large

$$P(|f(X_\alpha)^* - f(X_\alpha))_*| > \epsilon\}.$$

Let T be an arbitrary set. The space $l^\infty(T)$ is defined as the set of all uniformly bounded, real functions on T: all functions $z : T \to \mathbf{R}$ such that

$$||z||_T := \sup_{t \in T} |z(t)| < \infty$$

It is a metric space with respect to the *uniform distance* $d(z_1, z_2) = ||z_1 - z_2||_T$.

The space $l^\infty(T)$, or a suitable subspace of it, is a natural space for stochastic processes with bounded sample paths. A stochastic process is simply an indexed collection $\{X(t) : t \in T\}$ of random variables defined on the same probability space: every $X(t) : \Omega \to \mathbf{R}$ is a measurable map. If every sample path $t \mapsto X(t, \omega)$ is bounded, then a stochastic process yields a map $X : \Omega \to l^\infty(T)$. Sometimes, the sample paths have additional properties, such as measurability or continuity, and it may be fruitful to consider X as a map into a subspace of $l^\infty(T)$. If in either case the uniform metric is used, this does not make a difference for weak convergence of a net, but for measurability it can.

In most cases, a map $X : \Omega \to l^\infty(T)$ is a stochastic process. The small amount of measurability this gives may already be enough for asymptotic measurability. The special role played by the marginals $(X(t_1), \ldots, X(t_k))$, which are considered as maps into $\mathbf{R}^k$, is underlined by the following three results. Weak convergence in $l^\infty(T)$ can be characterized as asymptotic tightness plus convergence of marginals.

Lemma 6.4: Let $X_\alpha : \Omega_\alpha \to l^\infty(T)$ be asymptotically tight. Then it is asymptotically measurable if and only if $X_\alpha(t)$ is asymptotically measurable for every $t \in T$.

Lemma 6.5: Let X and Y be tight Borel measurable maps into $l^\infty(T)$. Then X and Y are equal in Borel law if and only if all corresponding marginals of X and Y are equal in law.

Theorem 6.3: Let $X_\alpha : \Omega_\alpha \to l^\infty(T)$ be arbitrary. Then X_α converges weakly to a tight limit if and only if X_α is asymptotically tight and the marginals $(X(t_1), \ldots, X(t_k))$ converge weakly to a limit for every finite subset $t_1, \ldots, t_k$ of T. If X_α is asymptotically tight and its marginals converge weakly to the marginals $(X(t_1), \ldots, X(t_k))$ of a stochastic process X, then there is a version of X with uniformly bounded sample paths and $X_\alpha \Rightarrow X$.

Proof: For the proof of both lemmas, consider the collection $\mathcal{F}$ of all functions $f : l^\infty(T) \to \mathbf{R}$ of the form

$$f(z) = g(z(t_1), \ldots, z(t_k)), \quad g \in C_b(\mathbf{R}^k), t_1, \ldots, t_k \in T, k \in \mathbf{N}.$$

This forms an algebra and a vector lattice, contains the constant functions, and separates points of $l^\infty(T)$. Therefore, the lemmas are corollaries of Lemmas 6.2 and 6.3, respectively. If X_α is asymptotically tight and marginals converge, then X_α is asymptotically measurable by the first lemma. By Prohorov's theorem, X_α is relatively compact $||z||_T = \sup_{t \epsilon T} |\mathbb{Z}(t)| < \infty$. To prove weak convergence, it suffices to show that all limit points are the same. This follows from marginal convergence and the second lemma.

Marginal convergence can be established by any of the well-known methods for proving weak convergence on Euclidean space. Tightness can be given a more concrete form, either through finite approximation or with the help of the Arzelà-Ascoli theorem. Finite approximation leads to the simpler of the two characterizations, but the second approach is perhaps of more interest, because it connects tightness to continuity of the sample paths $t \mapsto X_\alpha(t)$.

The idea of finite approximation is that for any $\epsilon > 0$, the index set T can be partitioned into finitely many subsets T_i such that the variation of the sample paths $t \mapsto X_\alpha(t)$ is less than ϵ on every one of the sets T_i. More precisely, it is assumed that for every $\epsilon, \eta > 0$, there exists a partition $T = \cup_{i=1}^k T_i$ such that

$$\limsup_\alpha P^*\Big(\sup_i \sup_{s,t \in T_i} |X_\alpha(s) - X_\alpha(t)| > \epsilon\Big) < \eta. \tag{6.1}$$

Clearly, under this condition, the asymptotic behavior of the process can be described within error margin ϵ, η by the behavior of the marginal $(X_\alpha(t_1), \dots, X_\alpha(t_k))$ for arbitrary fixed points $t_i \in T_i$. If the process can thus be reduced to a finite set of coordinates for any $\epsilon, \eta > 0$ and the nets or marginal distributions are tight, then the net X_α is asymptotically tight.

Theorem 6.4: A net $X_\alpha : \Omega_\alpha \to l^\infty(T)$ is asymptotically tight if and only if $X_\alpha(t)$ is asymptotically tight in $\mathbf{R}$ for every t and, for all $\epsilon, \eta > 0$, there exists a finite partition $T = \cup_{i=1}^k T_i$ such that (6.1) holds.

Proof: The necessity of the conditions follows easily from the next theorem. For instance, take the partition equal to disjointified balls of radius δ for a semi-metric on T as in the next theorem. We prove sufficiency.

For any partition, as in the condition of the theorem, the norm $||X_\alpha||_T$ is bounded by $max_i|X_\alpha(t_i)| + \epsilon$, with inner probability at least $1 - \eta$, if $t_i \in T_i$ for each i. Since a maximum of finitely many tight nets of real variables is tight, it follows that the net $||X_\alpha||_T$ is asymptotically tight in R.

Fix $\zeta > 0$ and a sequence $\epsilon_m \downarrow 0$. Take a constant M such that $\limsup P^*(||X_\alpha||_T > M) < \zeta$, and for each $\epsilon = \epsilon_m$ and $\eta = 2^{-m}\zeta$, take a partition $T = \cup_{i=1}^k T_i$ as in (61). For the moment, m is fixed and we do not let it appear in the notation. Let $z_1, \dots, z_p$ be the set of all functions in $l^\infty(T)$ that are constant on each T_i and take on only the values $0, \pm\epsilon_m, \dots, \pm[M/\epsilon_m]\epsilon_m$. Let K_m be the union of the p closed balls of radius ϵ_m around the z_i. Then, by construction, the two conditions

$$||X_\alpha||_T \le M \quad \text{and} \quad \sup_i \sup_{s,t\in T_i} \left|X_\alpha(s) - X_\alpha(t)\right| \le \epsilon_m$$

imply that $X_\alpha \in K_m$. This is true for each fixed m.

Let $K = \cap_{m=1}^\infty K_m$. Then K is closed and totally bounded (by construction of the K_m and because $\epsilon \downarrow 0$) and hence compact. Furthermore, for every $\delta > 0$, there is an m with $K^\delta \supset \cap_{m=1}^m K_i$. If not, then there would be a sequence z_m not in K^δ, but with $z_m \in \cap_{m=1}^m K_i$ for every m. This would have a subsequence contained in one of the balls making up K_1, a further subsequence eventually contained in one of the balls making up K_2, and so on. The "diagonal" sequence, formed by taking the first of the first subsequence, the second of the second subsequence, and so on would eventually be contained in a ball of radius ϵ_m for every m, hence Cauchy. Its limit would be in K, contradicting the fact that $d(z_m, K) \ge \delta$ for every m.

Conclude that if X_α is not in K^δ, then it is not in $\cap_{m=1}^m K_i$ for some fixed m. Then

$$\limsup P^*(X_\alpha \notin K^\delta) \le \limsup P^*\left(X_\alpha \notin \cap_{m=1}^m K_i\right) \le \zeta + \sum_{i=1}^m \zeta 2^{-m} < 2\zeta.$$

This concludes the proof of the theorem.

The second type of characterization of asymptotic tightness is deeper and relates the concept to asymptotic continuity of the sample paths. Suppose ρ is a semi-metric on T A net $X_\alpha : \Omega_\alpha \to l^\infty(T)$ is asymptotically uniformly ρ-equicontinuous in probability if for every $\epsilon, \eta > 0$ there exists a $\delta > 0$ such that

$$\limsup_{\alpha} P^*\Big(\sup_{\rho(s,t)<\delta} |X_\alpha(s) - X_\alpha(t)| > \epsilon \Big) < \eta.$$

Theorem 6.5: A net $X_\alpha : \Omega_\alpha \to l^\infty(T)$ is asymptotically tight if and only if $X_\alpha(t)$ is asymptotically tight in R for every t and there exists a semi-metric ρ on T such that (T, ρ) is totally bounded and X_α is asymptotically uniformly ρ-equicontinuous in probability. If, moreover, $X_\alpha \Rightarrow X$, then almost all paths $t \mapsto X(t, \omega)$ are uniformly ρ-continuous; and the semi-metric ρ can without loss of generality be taken equal to any semi-metric ρ for which this is true and (T, ρ) is totally bounded.

Proof: ($\Leftarrow$). The sufficiency follows from the previous theorem. First, take $\delta > 0$ sufficiently small so that the last displayed inequality is valid. Since T is totally bounded, it can be covered with finitely many balls of radius δ. Construct a partition of T by disjointifying these balls.

($\Rightarrow$). If X_α is asymptotically tight, then $g(X_\alpha)$ is asymptotically tight for every continuous map g; in particular, for each coordinate projection. Let $K_1 \subset K_2 \subset \ldots$ be compacts with $\liminf P_*(X_\alpha \in K_m^\epsilon)1 - 1/m$ for every $\epsilon > 0$. For every fixed m, define a semi-metric ρ_m on T by

$$\rho_m(s, t) = \sup_{z \in K_m} |z(s) - z(t)|, \quad s, t \in T.$$

Then (T, ρ_m) is totally bounded. Indeed, cover K_m by finitely many balls of radius η, centered at $z_1, \ldots, z_k$. Partition R^k into cubes of edge η, and for every cube pick at most one $t \in T$ such that $(z_1(t), \ldots, z_k(t))$ is in the cube. Since $z_1, \ldots, z_k$ are uniformly bounded, this gives finitely many points $t_1, \ldots, t_p$. Now the balls $\{t : \rho(t, t_i) < 3\eta\}$ cover T: t is in the ball around t_i for which $(z_1(t), \ldots, z_k(t))$ and $(z_1(t_i), \ldots, z_k(t_i))$ fall in the same cube. This follows because $\rho_m(t, t_i)$ can be bounded by $2\sup_{z \in K_m} \inf_i \|z - z_i\|_T + \sup_j |z_j(t_i) - z_j(t)|$. Next set

$$\rho(s, t) = \sum_{m=1}^{\infty} 2^{-m}(\rho_m(s, t) \wedge 1).$$

Fix $\eta > 0$. Take a natural number m with $2^{-m} < \eta$. Cover T with finitely many ρ_m-balls of radius η. Let $t_1, \ldots, t_p$ be their centers. Since $\rho_1 \le \rho_2 \le \ldots$, there is for every t a t_i with $\rho(t, t_i) \le \sum_{k=1}^m 2^{-k}\rho_k(t, t_i) + 2^{-m} < 2\eta$. Thus, (T, ρ) is totally bounded for ρ, too. It is clear from the definitions that $|z(s) - z(t)| \le \rho_m(s, t)$ for every $z \in K_m$ and that $\rho_m(s,t) \wedge 1 \le 2^m \rho(s,t)$. Also, if $\|z_0 - z\|_T < \epsilon$ for $z \in K_m$ then $|z_0(s) - z_0(t)| < 2\epsilon + |z(s) - z(t)|$ for any pair s, t. Deduce that

$$K_m^\epsilon \subset \Big\{z : \sup_{\rho(s,t)<2^{-m}\epsilon} |z(s) - z(t)| \le 3\epsilon\Big\}.$$

Thus, for given ϵ and m, and for $\delta < 2^{-m}\epsilon$,

$$\liminf_\alpha P_*\Big(\sup_{\rho(s,t)<\delta} \big|X_\alpha(s) - X_\alpha(t)\big| \le 3\epsilon\Big) \ge 1 - \frac{1}{m}.$$

Finally, if $X_\alpha \Rightarrow X$, then with notation as in the second part of the proof, $P(X \in K_m) \ge 1 - 1/m$; hence, X concentrates on $\cup_{m=1}^\infty K_m$. The elements of K_m are uniformly ρ_m-equicontinuous and hence also uniformly ρ-continuous. This yields the first statement. The set of uniformly continuous functions on a totally bounded, semi-metric space is complete and separable, so a map X that takes its values in this set is tight. Next if $X_\alpha \Rightarrow X$ and X is tight, the X_α is asymptotically tight and the compacts for asymptotical tightness can be chosen equal to the compacts for tightness of X. If X has uniformly continuous paths, then the latter compacts can be chosen within the space of uniformly continuous functions. Since a compact is totally bounded, every one of the compacts is necessarily uniformly equicontinuous. The combination of these facts proves the second statement.

6.2 Maximal inequalities and covering numbers

We derive a class of maximal inequalities that can be used to establish the asymptotic equicontinuity of the empirical process. Since the inequalities have much wider applicability, we temporarily leave the empirical framework.

Let ψ be a nondecreasing, convex function with $\psi(0) = 0$ and X a random variable. Then the Orlicz norm $\|X\|_\psi$ is defined as

$$\|X\|_\psi = \inf\Big\{C > 0 : E\psi\Big(\frac{|X|}{C}\Big) \le 1\Big\}.$$

Here the infimum over the empty set is ∞. Using Jensen's inequality, it is not difficult to check that this indeed defines a norm. The best-known examples of Orlicz norms are those corresponding to the functions $x \mapsto x^p$ for $p \ge 1$: the corresponding Orlicz norm is simply the L_p-norm

$$\|X\|_p = \big(E|X|^p\big)^{1/p}.$$

For our purpose, Orlicz norms of more interest are the ones given by $\psi_p(x) = e^{x^p} - 1$ for $p \geq 1$, which give much more weight to the tails of X. The bound $x^p \leq \psi_p(x)$ for all nonnegative x implies that $||X||_p \leq ||X||_{\psi_p}$ for each p. It is not true that the exponential Orlicz norms are all bigger than all L_p-norms. However, we have the inequalities

$$\begin{aligned} ||X||_{\psi_p} &\leq ||X||_{\psi_p}(\log 2)^{1/q-1/p}, \quad p \leq q \\ ||X||_p &\leq p!||X||_{\psi_1} \end{aligned}$$

Since for the present purposes fixed constants in inequalities are irrelevant, this means that a bound on an exponential Orlicz norm always gives a better result than a bound on an L_p-norm.

Any Orlicz norm can be used to obtain an estimate of the tail of a distribution. By Markov's inequality,

$$P(|X| > x) \leq P\Big(\psi(|X|/||X||_\psi) \geq \psi(x/||X||_\psi)\Big) \leq \frac{1}{\psi(x/||X||_\psi)}$$

For $\psi_p(x) = e^{x^p} - 1$, this leads to tail estimates $\exp(-Cx^p)$ for any random variable with a finite ψ_p-norm. Conversely, an exponential tail bound of this type shows that $||X||_{\psi_p}$ is finite.

Lemma 6.6: Let X be a random variable with $P(|X| > x) \leq Ke^{-Cx^p}$ for every x, for constants K and C, and for $p \geq 1$. Then its Orlicz norm satisfies $||X||_{\psi_p} \leq \big((1+K)/C\big)^{1/p}$.

Proof: By Fubini's theorem,

$$E\Big(e^{D|X|^p} - 1\Big) = E \int_0^{|X|^p} De^{Ds}\,ds = \int_0^\infty P(|X| > s^{1/p})De^{Ds}\,ds$$

Now insert the inequality on the tails of $|X|$ and obtain the explicit upper bound $KD/(C-D)$. This is less than or equal to 1 for $D^{-1/p}$ greater than or equal to $\big((1+K)/C\big)^{1/p}$. This completes the proof.

Next consider the ψ-norm of a maximum of finitely many random variables. Using the fact that $\max |X_i|^p \leq \sum |X_i|^p$, one easily obtains for the L_p-norms

$$\Big|\Big|\max_{1\leq i\leq m} X_i\Big|\Big|_p = \Big(E\max_{1\leq i\leq m} X_i^p\Big)^{1/p} \leq m^{1/p}\max_{1\leq i\leq m}||X_i||_p.$$

A similar inequality is valid for many Orlicz norms, in particular the exponential ones. Here, in the general case, the factor $m^{1/p}$ becomes $\psi^{-1}(m)$.

Lemma 6.7: Let ψ be a convex, nondecreasing, nonzero function with $\psi(0) = 0$ and $\limsup_{x,y\to\infty} \psi(x)\psi(y)/\psi(cxy) < \infty$ for some constant c. Then for any random variables $X_1, \ldots, X_n$,

$$\left\| \max_{1\le i\le m} X_i \right\|_\psi =\le K\psi^{-1}(m) \max_{1\le i\le m} \|X_i\|_p$$

for a constant K depending only on ψ.

Proof: For simplicity of notation, assume first that $\psi(x)\psi(y) \le \psi(cxy)$ for all $x, y \ge 1$. In that case, $\psi(x/y) \le \psi(cx)/\psi(y)$ for all $x \ge y \ge 1$. Thus, for $y \ge 1$ and any C,

$$\max \psi\left(\frac{|X_i|}{Cy}\right) \le \max\left[\frac{\psi(c|X_i|/C)}{\psi(y)} + \psi\left(\frac{|X_i|}{Cy}\right)1\left\{\frac{|X_i|}{Cy} < 1\right\}\right]$$
$$\le \sum \frac{\psi(c|X_i|/C)}{\psi(y)} + \psi(1).$$

Set $C = c\max \|X_i\|_\psi$, and take expectations to get

$$E\psi\left(\frac{\max|X_i|}{Cy}\right) \le \frac{m}{\psi(y)} + \psi(1).$$

When $\psi(1) \le 1/2$, this is less than or equal to 1 for $y = \psi^{-1}(2m)$, which is greater than 1 under the same condition. Thus,

$$\left\| \max_{1\le i\le m} X_i \right\|_\psi \le \psi^{-1}(2m)c \max \|X_i\|_\psi.$$

By the convexity of ψ and the fact that $\psi(0) = 0$, it follows that $\psi^{-1}(2m) \le 2\psi^{-1}(m)$. The proof is complete for every special ψ that meets the conditions made previously. For a general ψ, there are constants $\sigma \le 1$ and $\tau > 0$ such that $\phi(x) = \sigma\psi(\tau x)$ satisfies the conditions of the previous paragraph. Apply the inequality to ϕ, and observe that $\|X\|_\psi \le \|X\|_\phi/(\sigma\tau) \le \|X\|_\psi/\sigma$.

For the present purposes, the value of the constant in the previous lemma is irrelevant. The important conclusion is that the inverse of the ψ-function determines the size of the ψ-norm of a maximum in comparison to the ψ-norms of the individual terms. The ψ-norms grows slowest for rapidly increasing ψ. For $\psi(x) = e^{x^p} - 1$, the growth is at most logarithmic because

$$\psi_p^{-1}(m) = (\log(1+m))^{1/p}$$

The previous lemma is useless in the case of a maximum over infinitely many variables. However, such a case can be handled via repeated application of the lemma via a method known as chaining. Every random variable in the supremum is written as a sum of "little links," and the bound depends on the number and size of the little links needed. For a stochastic process $\{X_t : t \in T\}$, the number of links depends on the entropy of the index set for the semi-metric

$$d(s,t) = \|X_s - X_t\|_\psi.$$

The general definition of "metric entropy" is as follows.

Definition 6.3 (Covering numbers): Let (T, d) be an arbitrary semi-metric space. Then the covering number $N(\epsilon, d)$ is the minimal number of balls of radius ϵ needed to cover T. Call a collection of points ϵ-separated if the distance between each pair of points is strictly larger than ϵ. The packing number $D(\epsilon, d)$ is the maximum number of ϵ-separated points in T.

The corresponding entropy numbers are the logarithms of the covering and packing numbers, respectively.

For the present purposes, both covering and packing numbers can be used. In all arguments one can be replaced by the other through the inequalities

$$N(\epsilon, d) \le D(\epsilon, d) \le N\left(\frac{\epsilon}{2}, d\right).$$

Clearly, covering and packing numbers become bigger as $\epsilon \downarrow 0$. By definition, the semi-metric space T is totally bounded if and only if the covering and packing numbers are finite for every $\epsilon > 0$. The upper bound in the following maximal inequality depends on the rate at which $D(\epsilon, d)$ grows as $\epsilon \downarrow 0$, as measured through an integral criterion.

Theorem 6.6: Let ψ be a convex, nondecreasing, nonzero function with $\psi(0) = 0$ and $\limsup_{x,y\to\infty} \psi(x)\psi(y)/\psi(cxy) < \infty$, for some constant c. Let $\{X_t : t \in T\}$ be a separable stochastic process with

$$\|X_s - X_t\|_\psi \le C d(s,t), \quad \text{for every } s, t$$

for some semi-metric d on T and a constant C. Then, for any $\eta, \delta > 0$,

$$\left\| \sup_{d(s,t)<\delta} |X_s - X_t| \right\|_\psi \le K\left[\int_0^\eta \psi^{-1}(D(\epsilon, d))d\epsilon + \delta\psi^{-1}(D^2(\eta, d)) \right],$$

for a constant K depending on ψ and C only.

Corollary 6.1: The constant K can be chosen such that

$$\left\| \sup_{s,t} |X_s - X_t| \right\|_{\psi} \le K \int_0^{diamT} \psi^{-1}(D(\epsilon, d))d\epsilon,$$

where $diamT$ is the diameter of T.

Proof: Assume without loss of generality that the packing numbers and the associated "covering integral" are finite. Construct nested sets $T_0 \subset T_1 \subset \cdots \subset T$ such that every T_j is a maximal set of points such that $d(s, t) > \eta 2^{-j}$ for every $s, t \in T_j$, where "maximal" means that no point can be added without destroying the validity of the inequality. By the definition of packing numbers, the number of points in T_j is less than or equal to $D(\eta 2^{-j}, d)$.

"Link" every point $t_{j+1} \in T_{j+1}$ to a unique $t_j \in T_j$ such that $d(t_j, t_{j+1}) \le \eta 2^{-j}$. Thus, obtain for every t_{k+1} a chain $t_{k+1}, t_k, \dots, t_0$ that connects it to a point in T_0. For arbitrary points s_{k+1}, t_{k+1} in T_{k+1}, the difference in increments along their chains can be bounded by

$$\begin{aligned} |(X_{s_{k+1}} - X_{s_0}) - (X_{t_{k+1}} - X_{t_0})| &= \left| \sum_{j=0}^{k} (X_{s_{j+1}} - X_{s_j}) - \sum_{j=0}^{k} (X_{t_{j+1}} - X_{t_j}) \right| \\ &\le 2 \sum_{j=0}^{k} \max |X_u - X_v| \end{aligned}$$

where for fixed j the maximum is taken over all links (u, v) from T_{j+1} to T_j. Thus, the jth maximum is taken over at most $\#T_{j+1}$ links, with each link having a ψ-norm $\|X_u - X_v\|_{\psi}$ bounded by $Cd(u, v) \le C\eta 2^{-j}$. It follows with the help of Lemma 6.7 that, for a constant depending only on ψ and C,

$$\begin{aligned} \left\| \max_{s,t \in T_{k+1}} |(X_s - X_{s_0}) - (X_t - X_{t_0})| \right\|_{\psi} &\le K \sum_{j=0}^{k} \psi^{-1}(D(\eta 2^{-j-1}, d))\eta 2^{-j} \\ &\le 4K \int_0^{\eta} \psi^{-1}(D(\epsilon, d))d\epsilon. \end{aligned} \tag{6.2}$$

In this bound, s_0 and t_0 are the endpoints of the chains starting at s and t, respectively.

The maximum of the increments $|X_{s_{k+1}} - X_{t_{k+1}}|$ can be bounded by the maximum on the left side of (6.2) plus the maximum of the discrepancies $|X_{s_0} - X_{t_0}|$ at the end of the chains. The maximum of the latter discrepancies will be analyzed by a seemingly circular argument. For every pair of endpoints s_0, t_0 of chains starting at two points in

T_{k+1} within distance δ of each other, choose exactly one pair s_{k+1}, t_{k+1} in T_{k+1}, with $d(s_{k+1}, t_{k+1}) < \delta$, whose chains end at s_0, t_0. By definition of T_0, this gives at most $D^2(\eta, d)$ pairs. By the triangle inequality,

$$|X_{s_0} - X_{t_0}| \leq |(X_{s_0} - X_{s_{k+1}}) - (X_{t_0} - X_{t_{k+1}})| + |X_{s_{k+1}} - X_{t_{k+1}}|.$$

Take the maximum over all pairs of endpoints s_0, t_0 as above. Then the corresponding maximum over the first term on the right in the last display is bounded by the maximum in the left side of (6.2). It ψ-norm can be bounded by the right side of this equation. Combine this with (6.2) to find that

$$\Big\| \max_{s,t \in T_{k+1}, d(s,t) < \delta} |(X_s - X_{s_0}) - (X_t - X_{t_0})| \Big\|_\psi$$

$$\leq 8K \int_0^\eta \psi^{-1}(D(\epsilon, d)) d\epsilon + \| \max |X_{s_{k+1}} - X_{t_{k+1}}| \|_\psi.$$

Here the maximum on the right is taken over the pairs s_{k+1}, t_{k+1} in T_{k+1} uniquely attached to the pairs s_0, t_0 as above. Thus the maximum is over at most $D^2(\eta, d)$ terms, each of whose ψ-norm is bounded by δ. Its ψ-norm is bounded by $K\psi^{-1}(D^2(\eta, d))\delta$.

Thus, the upper bound given by the theorem is a bound for the maximum of increments over T_{k+1}. Let k tend to infinity to conclude the proof. The corollary follows immediately from the previous proof, after noting that, for η equal to the diameter of T, the set T_0 consists of exactly one point. In that case $s_0 = t_0$ for every pair s, t, and the increments at the end of the chains are zero. The corollary also follows from the theorem upon taking $\eta = \delta = diam T$ and noting that $D(\eta, d) = 1$, so that the second term in the maximal inequality can also be written $\delta\psi^{-1}(D(\eta, d))$. Since the function $\epsilon \mapsto \psi^{-1}(D(\epsilon, d))$ is decreasing, this term can be absorbed into the integral, perhaps at the cost of increasing the constant K.

Although the theorem gives a bound on the continuity modulus of the process, a bound on the maximum of the process will be needed. Of course, for any t_0,

$$\Big\| \sup_t |X_t| \Big\|_\psi \leq \|X_{t_0}\|_\psi + K \int_0^{diam T} \psi^{-1}(D(\epsilon, d)) d\epsilon.$$

Nevertheless, to state the maximal inequality in terms of the increments appears natural. The increment bound shows that the process X is continuous in ψ-norm, whenever the covering integral $\int_0^\eta \psi^{-1}(D(\epsilon, d)) d\epsilon$ converges for some $\eta > 0$. It is a small step to deduce the continuity of almost all sample paths from this inequality, but this is not needed at this point.

6.3 Sub-Gaussian inequalities

A standard normal variable has tails of the order $x^{-1}\exp\left(-\frac{x^2}{2}\right)$ and satisfies $P(|X| > x) \leq 2\exp\left(-\frac{x^2}{2}\right)$ for every x. In this section, we study random variables satisfying similar tail bounds.

Hoeffding's inequality asserts a "sub-Gaussian" tail bound for random variables of the form $X = \sum X_i$ with $X_1, \ldots, X_n$ i.i.d. with zero means and bounded range. The following special case of Hoeffding's inequality will be needed.

Theorem 6.7 (Hoeffding's inequality): Let $a_1, \ldots, a_n$ be constants and $\epsilon_1, \ldots, \epsilon_n$ be independent Rademacher random variables, i.e., with $P(\epsilon_i = 1) = P(\epsilon_i = -1) = 1/2$. Then

$$P\left(|\sum \epsilon_i a_i| > x\right) \leq 2e^{-\frac{x^2}{2||a||^2}},$$

for the Euclidean norm $||a||$. Consequently, $||\sum \epsilon_i a_i||_{\psi_2} \leq \sqrt{6}||a||$.

Proof: For any λ and Rademacher variable ϵ, one has $Ee^{\lambda\epsilon} = (e^{\lambda} + e^{-\lambda}) \leq e^{\lambda^2/2}$, where the last inequality follows after writing out the power series. Thus by Markov's inequality, for any $\lambda > 0$,

$$P\left(\sum \epsilon_i a_i > x\right) \leq e^{-\lambda x} Ee^{\lambda \sum_{i=1}^n a_i \epsilon_i} \leq e^{(\frac{\lambda^2}{2}||a||^2 - \lambda x)}.$$

The best upper bound is obtained for $\lambda = x/||a||^2$ and is the exponential in the probability bound of the lemma. Combination with a similar bound for the lower tail yields the probability bound. The bound on the ψ-norm is a consequence of the probability bound in view of Lemma 6.6.

A stochastic process is called sub-Gaussian with respect to the semi-metric d on its index set if

$$P(|X_s - X_t| > x) \leq 2e^{-\frac{x^2}{2d^2(s,t)}}, \qquad \text{for every } s, t \in T, x > 0$$

any Gaussian process is sub-Gaussian for the standard deviation semi-metric $d(s, t) = \sigma(X_s - X_t)$. Another example is Rademacher process

$$X_a = \sum_{i=1}^n a_i \epsilon_i, \qquad a \in R^n$$

for Rademacher variables $\epsilon_1, \ldots, \epsilon_n$. By Hoeffding's inequality, this is sub-Gaussian for the Euclidean distance $d(a, b) = ||a - b||$.

Sub-Gaussian processes satisfy the increment bound $\|X_s - X_t\|_{\psi_2} \le \sqrt{6}d(s,t)$. Since the inverse of the ψ_2-function is essentially the square root of the logarithm, the general maximal inequality leads for sub-Gaussian processes to a bound in terms of an entropy integral. Furthermore, because of the special properties of the logarithm, the statement can be slightly simplified.

Corollary 6.2: Let $\{X_t : t \in T\}$ be a separable sub-Gaussian process. Then for every $\delta > 0$,

$$E \sup_{d(s,t)\le\delta} |X_s - X_t| \le K \int_0^\delta \sqrt{\log D(\epsilon, d)}\, d\epsilon,$$

for a universal constant K. In particular, for any t_0,

$$E \sup_t |X_t| \le E|X_{t_0}| + K \int_0^\infty \sqrt{\log D(\epsilon, d)}\, d\epsilon.$$

Proof: Apply the general maximal inequality with $\psi_2(x) = e^{x^2} - 1$ and $\eta = \delta$. Since $\psi_2^{-1}(m) = \sqrt{\log(1+m)}$, we have $\psi_2^{-1}(D^2(\delta, d)) \le \sqrt{2}\psi_2^{-1}(D(\delta, d))$. Thus, the second term in the maximal inequality can first be replaced by $\sqrt{2}\delta\psi^{-1}(D(\eta, d))$ and next be incorporated in the first at the cost of increasing the constant. We obtain

$$\Big\| \sup_{d(s,t)\le\delta} |X_s - X_t| \Big\|_{\psi_2} \le K \int_0^\delta \sqrt{\log(1 + D(\epsilon, d))}\, d\epsilon.$$

Here $D(\epsilon, d) \ge 2$ for every ϵ that is strictly less than the diameter of T. Since $\log(1+m) \le 2\log m$ for $m \ge 2$, the 1 inside the logarithm can be removed at the cost of increasing K.

6.4 Symmetrization

Let $\epsilon_1, \ldots, \epsilon_n$ be i.i.d. Rademacher random variables. Instead of the empirical process

$$f \mapsto (P_n - P)f = \frac{1}{n}\sum_{i=1}^n (f(X_i) - Pf),$$

consider the symmetrized process

$$f \mapsto P_n^o f = \frac{1}{n}\sum_{i=1}^n \epsilon_i f(X_i),$$

where $\epsilon_1, \ldots, \epsilon_n$ are independent of $(X_1, \ldots, X_n)$. Both processes have mean function zero. It turns out that the law of large numbers or the central limit theorem for one of these processes holds if and only if the corresponding result is true for the other process. One main approach to proving empirical limit theorems is to pass from $P_n - P$ to P_n^o and next apply arguments conditionally on the original X's. The idea is that, for fixed $X_1, \ldots, X_n$, the symmetrized empirical measure is a Rademacher process, hence a sub-Gaussian process, to which Corollary 6.2 can be applied.

Thus, we need to bound maxima and moduli of the process $P_n - P$ by those of the symmetrized process. To formulate such bounds, we must be careful about the possible nonmeasurability of suprema of the type $||P_n - P||_{\mathcal{F}}$. The result will be formulated in terms of outer expectation, but it does not hold for every choice of an underlying probability space on which $X_1, \ldots, X_n$ are defined. Throughout this part, if outer expectations are involved, it is assumed that $X_1, \ldots, X_n$ are the coordinate projections on the product space $(\mathcal{X}^n, \mathcal{A}^n, P^n)$, and the outer expectations of functions $(X_1, \ldots, X_n) \mapsto h(X_1, \ldots, X_n)$ are computed for P^n. thus "independent" is understood in terms of a product probability space. If auxiliary variables, independent of the X's, are involved, as in the next lemma, we use a similar convention. In that case, the underlying probability space is assumed to be of the form $(\mathcal{X}^n, \mathcal{A}^n, P^n) \times (\mathcal{Z}, \mathcal{C}, Q)$ with $X_1, \ldots, X_n$ equal to the coordinate projections on the first n coordinates and the additional variables depending only on the $(n + 1)$st coordinate.

The following lemma will be used mostly with the choice $\Phi(x) = x$.

Lemma 6.8 (symmetrization): For every nondecreasing, convex $\Phi : R \to R$ and class of measurable functions $\mathcal{F}$,

$$E^*\Phi\Big(||P_n - P||_{\mathcal{F}}\Big) \le E^*\Phi\Big(2||P_n^0||_{\mathcal{F}}\Big),$$

where the outer expectations are computed as indicated in the preceding paragraph.

Proof: Let $Y_1, \ldots, Y_n$ be independent copies of $X_1, \ldots, X_n$, defined formally as the coordinate projections on the last n coordinates in the product space $(\mathcal{X}^n, \mathcal{A}^n, P^n) \times (\mathcal{Z}, \mathcal{C}, Q) \times (\mathcal{X}^n, \mathcal{A}^n, P^n)$. The outer expectations in the statement of the lemma are unaffected by this enlargement of the underlying probability space, because coordinate projections are perfect maps. For fixed values $X_1, \ldots, X_n$,

$$\begin{aligned} ||P_n - P||_{\mathcal{F}} &= \sup_{f \in \mathcal{F}} \frac{1}{n}\Big|\sum_{i=1}^{n} (f(X_i) - Ef(Y_i))\Big| \\ &\le E_Y^* \sup_{f \in \mathcal{F}} \frac{1}{n}\Big|\sum_{i=1}^{n} (f(X_i) - f(Y_i))\Big|, \end{aligned}$$

where E_Y^* is the outer expectation with respect to $Y_1, \ldots, Y_n$ computed for P^n for given, fixed values of $X_1, \ldots, X_n$. Combination with Jensen's inequality yields

$$\Phi\Big(||P_n - P||_{\mathcal{F}}\Big) \le E_Y\Phi\left(\Big\|\frac{1}{n}\sum_{i=1}^{n} (f(X_i) - f(Y_i))\Big\|_{\mathcal{F}}^{*Y}\right),$$

where $*Y$ denotes the minimal measurable majorant of the supremum with respect to $Y_1, \ldots, Y_n$, still with $X_1, \ldots, X_n$ fixed. Because Φ is nondecreasing and continuous, the $*Y$ inside Φ can be moved to E_Y^*. Next take the expectation with respect to $X_1, \ldots, X_n$ to get

$$E^*\Phi\left(\|P_n - P\|_{\mathcal{F}}\right) \le E_X^* E_Y^* \Phi\left(\left\|\frac{1}{n}\sum_{i=1}^n (f(X_i) - f(Y_i))\right\|_{\mathcal{F}}\right).$$

Here the repeated outer expectation can be bounded above by the joint outer expectation E^* by Fubini's theorem.

Adding a minus sign in front of a term $(f(X_i) - f(Y_i))$ has the effect of exchanging X_i and Y_i. By construction of the underlying probability space as a product space, the outer expectation of any function $f(X_1, \ldots, X_n, Y_1, \ldots, Y_n)$ remains unchanged under permutations of its $2n$ arguments, hence the expression

$$E^*\Phi\left(\left\|\frac{1}{n}\sum_{i=1}^n e_i(f(X_i) - f(Y_i))\right\|_{\mathcal{F}}\right)$$

is the same for any n-tuple $(e_1, \ldots, e_n) \in \{-1, 1\}^n$. Deduce that

$$E^*\Phi\left(\|P_n - P\|_{\mathcal{F}}\right) \le E_\epsilon E_{X,Y}^* \Phi\left(\left\|\frac{1}{n}\sum_{i=1}^n \epsilon_i(f(X_i) - f(Y_i))\right\|_{\mathcal{F}}\right).$$

Use the triangle inequality to separate the contributions of the X's and the Y's and next use the convexity of Φ and triangle inequality to bound the previous expression by

$$\frac{1}{2}E_\epsilon E_{X,Y}^* \Phi\left(2\left\|\frac{1}{n}\sum_{i=1}^n \epsilon_i f(X_i)\right\|_{\mathcal{F}}\right) + \frac{1}{2}E_\epsilon E_{X,Y}^* \Phi\left(2\left\|\frac{1}{n}\sum_{i=1}^n \epsilon_i f(Y_i)\right\|_{\mathcal{F}}\right).$$

By perfectness of coordinate projections, the expectation $E_{X,Y}^*$ is the same as E_X^* and E_Y^* in the two terms, respectively. Finally, replace the repeated outer expectations by a joint outer expectation. This completes the proof.

The symmetrization lemma is valid for any class $\mathcal{F}$. In the proofs of Glivenko-Cantelli and Donsker theorems, it will be applied not only to the original set of functions of interest, but also to several classes constructed from such a set $\mathcal{F}$. The next step in these proofs is to apply a maximal inequality to the right side of the lemma, conditionally on $X_1, \ldots, X_n$. At that point, we need to write the joint outer expectation as the repeated expectation $E_X^* E_\epsilon$, where the indices X and ϵ mean expectation over X and ϵ conditionally on remaining variables. Unfortunately, Fubini's theorem is not valid for outer expectations. To overcome this problem, it is assumed that the integrand in the right side of the lemma is jointly measurable in $(X_1, \ldots, X_n, \epsilon_1, \ldots, \epsilon_n)$.

Since the Rademacher variables are discrete, this is the case if and only if the maps

$$(X_1, \dots, X_n) \mapsto \left\| \sum_{i=1}^{n} e_i f(X_i) \right\|_{\mathcal{F}} \tag{6.3}$$

are measurable for every n-tuple $(e_1, \dots, e_n) \in \{-1, 1\}^n$. For the intended application of Fubini's theorem, it suffices that this is the case for the completion of $(\mathcal{X}^n, \mathcal{A}^n, P^n)$.

Definition 6.4 (measurable class): A class $\mathcal{F}$ of measurable functions $f : \mathcal{X} \to R$ on a probability space $(\mathcal{X}, \mathcal{A}, P)$ is called a P-measurable class if the function (6.3) is measurable on the completion of $(\mathcal{X}^n, \mathcal{A}^n, P^n)$ for every n and every vector $(e_1, \dots, e_n) \in R^n$.

6.4.1 Glivenko-Cantelli theorems

In this section, we prove two types of Glivenko-Cantelli theorems. The first theorem is the simplest and is based on entropy with bracketing. Its proof relies on finite approximation and the law of large numbers for real variables. The second theorem uses random L_1-entropy numbers and is proved through symmetrization followed by a maximal inequality.

Definition 6.5 (Covering numbers): The covering number $N(\epsilon, \mathcal{F}, ||\cdot||)$ is the minimal number of balls $\{g : ||g-f|| < \epsilon\}$ of radius ϵ needed to cover the set $\mathcal{F}$. The centers of the balls need not belong to $\mathcal{F}$, but they should have finite norms. The entropy (without bracketing) is the logarithm of the covering number.

Definition 6.6 (bracketing numbers): Given two functions l and u, the bracket $[l, u]$ is the set of all functions f with $l \leq f \leq u$. An ϵ-bracket is a bracket $[l, u]$ with $||u - l|| < \epsilon$. The bracketing number $N_{[]}(\epsilon, \mathcal{F}, ||\cdot||)$ is the minimum number of ϵ-brackets needed to cover $\mathcal{F}$. The entropy with bracketing is the logarithm of the bracketing number. In the definition of the bracketing number, the upper and lower bounds u and l of the brackets need not belong to $\mathcal{F}$ themselves but are assumed to have finite norms.

Theorem 6.8: Let $\mathcal{F}$ be a class of measurable functions such that $N_{[]}(\epsilon, \mathcal{F}, L_1(P)) < \infty$ for every $\epsilon > 0$. Then $\mathcal{F}$ is Glivenko-Cantelli.

Proof: Fix $\epsilon > 0$. Choose finitely many ϵ-brackets $[l_i, u_i]$ whose union contains $\mathcal{F}$ and such that $P(u_i - l_i) < \epsilon$ for every i. Then for every $f \in \mathcal{F}$, there is a bracket such that

$$(P_n - P)f \leq (P_n - P)u_i + P(u_i - f) \leq (P_n - P)u_i + \epsilon$$

Consequently,

$$\sup_{f\in\mathcal{F}}(P_n - P)f \leq \max_i (P_n - P)u_i + \epsilon.$$

The right side converges almost surely to ϵ by the strong law of large numbers for real variables. Combination with a similar argument for $\inf_{f\in\mathcal{F}}(P_n - P)f$ yields that $\limsup \|P_n - P\|_{\mathcal{F}}^* \leq \epsilon$ almost surely, for every $\epsilon > 0$. Take a sequence $\epsilon_m \downarrow 0$ to see that the limsup must actually be zero almost surely. This completes the proof.

An envelope function of a class $\mathcal{F}$ is any function $x \mapsto F(x)$ such that $|f(x)| \leq F(x)$, for every x and f. The minimal envelope function is $x \mapsto sup_f|f(x)|$. It will usually be assumed that this function is finite for every x.

Theorem 6.9: Let $\mathcal{F}$ be a P-measurable class of measurable functions with envelope F such that $P^*F < \infty$. Let $\mathcal{F}_M$ be the class of functions $f1\{F \leq M\}$ when f ranges over $\mathcal{F}$. If $\log N(\epsilon, \mathcal{F}_M, L_1(P_n)) = o_P^*(n)$ for every ϵ and $M > 0$, then $\|P_n - P\|_{\mathcal{F}}^* \to 0$ both almost surely and in mean. In particular, $\mathcal{F}$ is Glivenko-Cantelli.

Proof: By the symmetrization lemma, measurability of the class $\mathcal{F}$, and Fubini's theorem,

$$\begin{aligned} E^*\|P_n - P\|_{\mathcal{F}} &\leq 2E_X E_\epsilon \left\| \frac{1}{n}\sum_{i=1}^n \epsilon_i f(X_i) \right\|_{\mathcal{F}} \\ &\leq 2E_X E_\epsilon \left\| \frac{1}{n}\sum_{i=1}^n \epsilon_i f(X_i) \right\|_{\mathcal{F}_M} + 2P^*F\{F > M\} \end{aligned}$$

by the triangle inequality, for every $M > 0$. For sufficiently large M, the last term is arbitrarily small. To prove convergence in mean, it suffices to show that the first term converges to zero for fixed M. Fix $X_1, \ldots, X_n$. If $\mathcal{G}$ is an ϵ-net in $L_1(P_n)$ over $\mathcal{F}_M$, then

$$E_\epsilon \left\| \frac{1}{n}\sum_{i=1}^n \epsilon_i f(X_i) \right\|_{\mathcal{F}_M} \leq E_\epsilon \left\| \frac{1}{n}\sum_{i=1}^n \epsilon_i f(X_i) \right\|_{\mathcal{G}} + \epsilon.$$

The cardinality of $\mathcal{G}$ can be chosen equal to $N(\epsilon, \mathcal{F}_M, L_1(P_n))$. Bound the L_1-norm on the right by the Orlicz-norm for $\psi_2(x) = \exp(x^2) - 1$, and use the maximal inequality Lemma 6.7 to find that the last expression does not exceed a multiple of

$$\sqrt{1 + \log N(\epsilon, \mathcal{F}_M, L_1(P_n))} \sup_{f\in\mathcal{G}} \left\| \frac{1}{n}\sum_{i=1}^n \epsilon_i f(X_i) \right\|_{\psi_2|X} + \epsilon, \tag{6.4}$$

where the Orlicz norms $\|\cdot\|_{\psi_2|X}$ are taken over $\epsilon_1, \ldots, \epsilon_n$ with $X_1, \ldots, X_n$ fixed. By Hoeffding's inequality, they can be bounded by $\sqrt{6/n}(\mathbb{P}_n f^2)^{1/2}$, which is less than $\sqrt{6/n}M$. Thus, the last displayed expression is bounded by

$$\sqrt{1 + \log N(\epsilon, \mathcal{F}_M, L_1(\mathbb{P}_n))}\sqrt{\frac{6}{n}}M + \epsilon \to_{P^*} \epsilon$$

It has been shown that the left side of (6.4) converges to zero in probability. Since it is bounded by M, its expectation with respect to $X_1, \ldots, X_n$ converges to zero by the dominated convergence theorem. This concludes the proof that $\|\mathbb{P}_n - P\|^*_{\mathcal{F}}$ in mean. That it also converges almost surely follows from the fact that the sequence $\|\mathbb{P}_n - P\|^*_{\mathcal{F}}$ is a reverse martingale with respect to a suitable filtration.

Fact: Let $\mathcal{F}$ be class of measurable functions with envelope F such that $P^*F < \infty$. Define $\sum_n$ be the σ-field generated by measurable functions $h : X^\infty \to \mathbb{R}$ that are permutation symmetric in first arguments. Then

$$E(\|\mathbb{P}_n - P\|^*_{\mathcal{F}} | \Sigma_{n+1}) \geq \|\mathbb{P}_{n+1} - P\|^*_{\mathcal{F}} \quad \text{a.s.}$$

6.4.2 Donsker theorems

Uniform entropy: In this section, the weak convergence of the empirical process will be established under the condition that the envelope function F be square integrable, combined with the uniform entropy bound

$$\int_0^\infty \sup_Q \sqrt{\log N(\epsilon\|F\|_{Q,2}, \mathcal{F}, L_2(Q))}d\epsilon < \infty. \tag{6.5}$$

Here the supremum is taken over all finitely discrete probability measures Q on $(\mathcal{X}, \mathcal{A})$ with $\|F\|^2_{Q,2} = \int F^2 dQ > 0$. These conditions are by no means necessary, but they suffice for many examples. The finiteness of the previous integral will be referred to as the uniform entropy condition.

Theorem 6.10: Let $\mathcal{F}$ be a class of measurable functions that satisfies the uniform entropy bound (6.5). Let the class $\mathcal{F}_\delta = \{f - g : f, g \in \mathcal{F}, \|f - g\|_{P,2} < \delta\}$ and $\mathcal{F}^2_\infty$ be P-measurable for every $\delta > 0$. If $P^*F^2 < \infty$, then $\mathcal{F}$ is P-Donsker.

Proof: Let $\delta_n \downarrow 0$ be arbitrary. By Markov's inequality and the symmetrization lemma,

$$P^*(\|\mathbb{G}_n\|_{\mathcal{F}_{\delta_n}} > x) \leq \frac{2}{x}E^*\left\|\frac{1}{\sqrt{n}}\sum_{i=1}^n \epsilon_i f(X_i)\right\|_{\mathcal{F}_{\delta_n}}.$$

Since the supremum in the right-hand side is measurable by assumption, Fubini's theorem applies and the outer expectation can be calculated as $E_X E_\epsilon$. Fix $X_1, \ldots, X_n$. By Hoeffding's inequality, the stochastic process $f \mapsto \{n^{-1/2} \sum_{i=1}^n \epsilon_i f(X_i)\}$ is sub-Gaussian for the $L_2(P_n)$-seminorm

$$\|f\|_n = \sqrt{\frac{1}{n} \sum_{i=1}^n f^2(X_i)}.$$

Use the second part of the maximal inequality Corollary 6.2 to find that

$$E_\epsilon \left\| \frac{1}{\sqrt{n}} \sum_{i=1}^n \epsilon_i f(X_i) \right\|_{\mathcal{F}_{\delta_n}} \leq \int_0^\infty \sqrt{\log N(\epsilon, \mathcal{F}_{\delta_n}, L_2(P_n))} d\epsilon.$$

For large values of ϵ, the set $\mathcal{F}_{\delta_n}$ fits in a single ball of radius ϵ around the origin, in which case the integrand is zero. This is certainly the case for values of ϵ larger than θ_n, where

$$\theta_n^2 = \sup_{f \in \mathcal{F}_{\delta_n}} \|f\|_n^2 = \left\| \frac{1}{n} \sum_{i=1}^n f^2(X_i) \right\|_{\mathcal{F}_{\delta_n}}.$$

Furthermore, covering numbers of the class $\mathcal{F}_\delta$ are bounded by covering numbers of $\mathcal{F}_\infty = \{f - g : f, g \in \mathcal{F}\}$. The latter satisfy $N(\epsilon, \mathcal{F}_\infty, L_2(Q)) \leq N^2(\epsilon/2, \mathcal{F}, L_2(Q))$ for every measure Q.

Limit the integral in (6.5) to the interval $(0, \theta_n)$, make a change of variables, and bound the integrand to obtain the bound

$$\int_0^{\theta_n/\|F\|_n} \sup_Q \sqrt{\log N(\epsilon \|F\|_{Q,2}, \mathcal{F}, L_2(Q))} d\epsilon \|F\|_n.$$

Here the supremum is taken over all discrete probability measures. The integrand is integrable by assumption. Furthermore, $\|F\|_n$ is bounded below by $\|F_*\|_n$, which converges almost surely to its expectation, which may be assumed positive. Use the Cauch-Schwarz inequality and the dominated convergence theorem to see that the expectation of this integral converges to zero provided $\theta_n \to_{P*} 0$. This would conclude the proof of asymptotic equicontinuity.

Since $\sup\{Pf^2 : f \in \mathcal{F}_{\delta_n}\} \to 0$ and $\mathcal{F}_{\delta_n} \subset \mathcal{F}_\infty$, it is certainly enough to prove that

$$\|P_n f^2 - Pf^2\|_{\mathcal{F}_\infty} \to_{P*} 0.$$

This is a uniform law of large numbers for the class $\mathcal{F}_\infty^2$. This class has integrable envelope $(2F)^2$ and is measurable by assumption. For any pair f, g of functions in $\mathcal{F}_\infty$,

$$P_n|f^2 - g^2| \leq P_n|f - g|4F \leq \|f - g\|_n \|4F\|_n.$$

It follows that the covering number $N(\epsilon||2F||_n^2, \mathcal{F}_\infty^2, L_1(P_n))$ is bounded by the covering number $N(\epsilon||F||_n, \mathcal{F}_\infty, L_2(P_n))$. By assumption, the latter number is bounded by a fixed number, so its logarithm is certainly $o_P^*(n)$, as required for the uniform law of large numbers, Theorem 6.8. This concludes the proof of asymptotic equicontinuity.

Finally, we show that $\mathcal{F}$ is totally bounded in $L_2(P)$. By the result of the last paragraph, there exists a sequence of discrete measures P_n with $||(P_n - P)f^2||_{\mathcal{F}_\infty}$ converging to zero. Take n sufficiently large so that the supremum is bounded by ϵ^2. by assumption, $N(\epsilon, \mathcal{F}, L_2(P_n))$ is finite. Any ϵ-net for $\mathcal{F}$ in $L_2(P_n)$ is a $\sqrt{2}\epsilon$-net in $L_2(P)$. This completes the proof.

6.5 Lindberg-type theorem and its applications

In this section, we want to show how the methods developed earlier of symmetrization and entropy bounds can be used to prove a limit theorem for sums of independent stochastic processes $Z_{ni}, f \in \mathcal{F}$ with bounded sample paths indexed by an arbitrary set $\mathcal{F}$. We need some preliminarily results for this purpose.

Lemma 6.5.1: Let $Z_1, Z_2, \ldots Z_n$ be independent stochastic processes with mean 0. With $|| \, ||_{\mathcal{F}}$ denoting the sup norm on $\mathcal{F}$, we get

$$E^*\Phi(\frac{1}{2}||\sum_{i=1}^n \epsilon_i Z_i||_{\mathcal{F}}) \leq E^*\Phi(||\sum_{i=1}^n Z_i||_{\mathcal{F}}) \leq E^*\Phi(2||\sum_{i=1}^n \epsilon_i(Z_i - \mu_i)||_{\mathcal{F}})$$

for every non-decreasing, convex $\Phi : \mathbb{R} \to \mathbb{R}$ and arbitrary functions $\mu_i : \mathcal{F} \to \mathbb{R}$.

Proof: The inequality on the right can be proved using techniques as in the proof of symmetrization Lemma and is left to the reader. For the inequality on the left, let $Y_1, Y_2, \ldots Y_n$ be independent copy of $Z_1, \ldots Z_n$ defined on $(\prod_{i=1}^n(\mathcal{X}_i, a_i, P_i)x(\mathcal{Z}, \mathcal{G}, \mathcal{Q})x\prod_1^n(\mathcal{X}_i, a_i, P_i)$ (the ϵ_i's are defined on $(\mathcal{Z}, \mathcal{G}, \mathcal{Q})$) and depend on the last n co-ordinates exactly as $Z_1, \ldots Z_n$ depend on the first n coordinates. Since $EY_i(f) = 0$, the LHS of the above expression is the average of

$$E^*\Phi(||\frac{1}{2}\sum_{i=1}^n \epsilon_i[Z_i(f) - EY_i(f)||_{\mathcal{F}}),$$

where $(e_1 \ldots e_n)$ range over $\{-1, 1\}^n$

By Jenssen's inequality

$$\begin{aligned}&E_{Z,Y}^*\Phi(||\tfrac{1}{2}\textstyle\sum_{i=1}^n e_i[Z_i(f) - Y_i(f)||_{\mathcal{F}})\\ &\leq E_{Z,Y}^*\Phi(||\tfrac{1}{2}\textstyle\sum_{i=1}^n[Z_i(f) - Y_i(f)]||_{\mathcal{F}}).\end{aligned}$$

Apply triangle inequality and convexity of Φ to complete the proof.

Lemma 6.5.2: For arbitrary stochastic processes $Z_1, \ldots Z_n$, arbitrary functions $\mu_1, \ldots \mu_n$: $\mathcal{F} \to \mathbb{R}$, and $x > 0$

$$\beta_n(x) P^*(||\sum_{i=1}^{n} Z_i||_{\mathcal{F}} > x)$$

$$\leq 2P^*(4||\sum_{i=1}^{n} \epsilon_i(Z_i - \mu_i)||_{\mathcal{F}} > x)$$

and $\beta_n(x) \leq inf_f P(1 \sum_{i=1}^{n} Z_i(f)| < x/2)$. In particular, this is true for i.i.d. mean 0 processes and $\beta_n(x) = 1 - 4\frac{n}{x^2} sup_f var(Z_1(f))$.

Proof: Let $Y_1, \ldots Y_n$ be independent copy of $Z_1, Z_2, \ldots Z_n$ as defined above. If $|| \sum_1^n Z_i||_{\mathcal{F}} > x$, then there is some f for which

$$|\sum_1^n Z_i(f)| > x.$$

Fix a realization of $Z_1, \ldots Z_n$ and corresponding f. For this realization,

$$\beta \leq P_Y^*(1 \sum_{i=1}^{n} Y_i(f)| < x/2)$$

$$\leq P_Y^*(1 \sum_{i=1}^{n} Y_i(f) - \sum_{i=1}^{1} Z_i(f)| > x/2)$$

$$\leq P_Y^*(||\sum_{i=1}^{n}(Y_i - Z_i)||_{\mathcal{F}} > x/2).$$

The far left and far right sides do not depend on a particular f, and the inequality between them is valid on the set

$$\{||\sum_1^n Z_i||_{\mathcal{F}} > x\}.$$

Integrate both sides with respect to $Z_1, \ldots Z_n$ over the set to obtain

$$\beta P^*(||\sum_{i=1}^{n} Z_i|| > x) \leq P_Z^* P_Y^*(||\sum_{i=1}^{n}(Y_i - Z_i)| > x/2).$$

By symmetry, the RHS equals

$$E_\epsilon P_Z^* P_Y^*(||\sum_{i=1}^{n} \epsilon_i(Y_i - Z_i)||_{\mathcal{F}} > x/2).$$

In view of triangle inequality, this is less than or equal to

$$2P^*(||\sum_1^n \epsilon_i(Y_i - \mu_i)||_{\mathcal{F}} > x/4).$$

Assuming i.i.d. $Z_1, \ldots Z_n$, one can derive the inequality from Chebyshev inequality.

Proposition 6.5.3: Let $0 < p < \infty$ and $X_1, \ldots X_n$ be independent stochastic processes indexed by T. Then there exist constants C_p and $0 < \mu_p < 1$ such that

$$E^* \max_{k \le n} ||S_k||^p \le C_p(E^* \max_{k \le n} ||X_k||^p + F^{-1}(\mu_p)^P),$$

where F^{-1} is quantile function of $\max_{k \le n} ||S_k||^*$. If $X_1, \ldots X_n$ are symmetric, then there exist K_p and ($0 < v_p < 1$) such that

$$E^* ||S_n||^p \le K_p(E^* \max_{k \le n} ||x_k||^p + G^{-1}(v_p)^p)$$

where G^{-1} is quantile function of $||S_n||^*$. For each $p \le 1$, the last inequality has a version for mean zero processes.

Proof: This is based on the following fact due to Höffman-Jørgensen.

Fact: Let $X_1, \ldots X_n$ be independent stochastic processes indexed by an arbitrary set. Then for $\lambda, \eta > 0$

$$(1) \quad P^*(\max_{k \le n} ||S_k|| > 3\lambda + \eta) \le P^*(\max_{isK \le n} ||S_k|| > \lambda)^2 + P^*(\max_{k \le n} ||x_k|| > \eta).$$

If $X_1, \ldots X_n$ are independent symmetric, then

$$(2) \quad P^*(||S_n|| > 2\lambda + \eta) \le 4P^*(||S_n|| > \lambda)^2 + P^*(\max_{k \le n} ||x_k|| > \eta).$$

Proof of Proposition:
Take $\lambda = \eta = t$ in the above inequality (1) to find that for $t > 0$

$$\begin{aligned} E^* \max_{k \le n} ||S_k||^p &\le 4^p \int P^*(\max_{k \le n} ||S_k|| > 4t) d(t^p) \\ &\le (4t)^p + 4^p \int_t^\infty P^*(||S_k|| > t)^2 d(t^p) + 4^p \int_t^\infty P^*(\max_{k \le n} ||X_k||^* > t d(t^p) \\ &\le (4t)^p + 4^p P^*(\max_{k \le n} ||S_k|| > t) E^* \max_{k \le n} ||S_n||^p + 4^p E^* \max_{k \le n} ||X_k||^p. \end{aligned} \quad (6.6)$$

Choose t such that $4^p P^*(\max_{k\leq n} ||S_k|| > t) < 1/2$ we get the inequality. Similar arguments using inequality (2) gives the second inequality.

The inequality for mean-zero processes follows using symmetrization using the inequality above. Then one can using desymmetrization as by Jensen inequality $E^*||S_n||^p$ is bounded by $E^*(||S_n - T_n||^p)$ if T_n is sum of n independent copies of $X_1, \ldots X_n$.

For each n, let $Z_1, \ldots Z_n$, m_n be independent stochastic processes indexed by a common semi-metric space $(\mathcal{F}, \mathcal{P})$. These processes are defined on a product probability space of dimension m_n. Define random semi-metric

$$d_n^2(f, g) = \sum_{i=1}^{m_n} (Z_{ni}(f) - Z_{ni}(g))^2.$$

The condition below in Theorem 6.5.1 is random entropy condition.

Theorem 6.5.1 (Lindberg): For each n, let $Z_1, \ldots Z_n$, m_n be independent stochastic processes indexed by $(\mathcal{F}, \mathcal{P})$ (a totally bounded semi-metric space).

Assume

(a) $\sum_1^{m_n} E^*(||Z_{ni}||_{\mathcal{F}}^2 \{||Z_{ni}||_{\mathcal{F}} > \eta) \to 0$ for each $\eta > 0$

(b) For each $(f, g)\epsilon$, $\mathcal{F} \otimes \mathcal{F}$
$\sum_1^{m_n} E(Z_{ni}(f), Z_{ni}(g)) \to C(f, g)$ finite

(c) $\sup_{p(f,g)<\delta_n} \sum_1^n E(Z_{ni}(f) - Z_{ni}(g))^2 \to$ for $\delta_n \downarrow 0$

(d) $\int_0^{\delta_n} \sqrt{\log N(\epsilon, f, dn)} d\epsilon \overset{P^*}{\to} 0$ for $\delta_n \downarrow 0$

Then the sequence $\sum_{i=1}^n (Z_{ni} - EZ_{ni})$ is asymptotically p-equicontinuous. It converges in distrubtion in $l^\infty(\mathcal{F})$.

Proof: The condition (a) implies Lindberg condition for $\{Z_{ni}(f), i = 1, 2 \ldots m_n\}$ and using (b) we get marginal distributions of sum converge to a Gaussian process.

Set $Z_{ni}^0 = Z_{ni} - EZ_{ni}$ and let $\delta_n \downarrow 0$ arbitrary. For fixed t and n large, Chebychev inequality and (c) give the bound

$$P(|\sum_{i=1}^n (Z_{ni}^0(f) - Z_{ni}^0(g))| > t/2) < 1/2$$

for f, g with $P(f, g) < \delta_n$. By Lemma 6.8, for sufficiently large n,

$$P^*(\sup_{p(f,g)<\delta_n} |\sum_1^{m_n} (Z_{ni}^0(f) - Z_{ni}^0(g))| > t)$$

$$\leq 4P(\sup_{p(f,g)<\delta_n} |\sum_1^{m_n} \epsilon_i(Z_{ni}(f) - Z_{ni}(g))| > t/4).$$

For fixed values of processes $Z_1, \ldots Z_n$, m_n define $A \leq \mathbb{R}^{m_n}$ as a set of all vectors.

$$(Z_{ni}(f) - Z_{ni}(g), \ldots Z_{nm_n}(f) - Z_{nm_n}(g))$$

when f, g are in the set $\{(f,g)\epsilon\mathcal{F}x\mathcal{F}, P(f,g) < \delta_n\}$. By Höffding inequality (Theorem 6.7), the stochastic process $\{\sum \epsilon_i a_i, a\epsilon A\}$ is Gaussian for Euclidean metric on A_n. By Corollary 6.9

$$E_\epsilon \sup_{p(f,g)<\delta_n} |\sum_{i=1}^{m_n} \epsilon_i(Z_{ni}(f) - Z_{ni}(g))|$$
$$\leq \int_0^\infty \sqrt{\log N(\epsilon, A_n, ||\ ||)} d\epsilon,$$

where $||\ ||$ is the Euclidean norm. The integrand can be bounded using $N(\epsilon, A_n, ||\ ||) \leq N^2(\epsilon/2, \mathcal{F}, dn)$. If, in addition,

$$\Theta_n^2 = \sup_{a\epsilon A_n} \sum_{i=1}^n a_i^2 = ||\sum_{i=1}^{m_n} a_i^2||_{A_n},$$

then for $\epsilon > \Theta_n$, the set A_n fits in the ball of radius ϵ around the origin and the integrand vanishes. From entropy condition (d) and this, we conclude that the integral converges to zero in outer probability if $\Theta_n \to 0$ in probability. Under the measurability assumption, this gives equicontinuity of $\sum_1^{m_n} Z_{ni}^0$.

By the Lindberg condition, there exist a sequence $\eta_n \downarrow 0$ such that

$$E^* ||\sum a_i^2\{||Z_{ni}|| > \eta_n\}||_{A_n} \to 0.$$

Thus, without loss of generality, we can assume $||Z_{ni}||_{\mathcal{F}} \leq \eta_n$ to show $\Theta_n \overset{P^*}{\to} 0$.

Fix $Z_{n1}, \ldots Z_{nm_n}$ and take ϵ-net b_n for A_n with Euclidean norm. For every $a\epsilon A_n$, there is a $b\epsilon B_n$ with

$$|\sum_i^n \epsilon_i a_i^2| = |\sum \epsilon_i(a_i - b_i)^2 + 2\sum(a_i - b_i)\epsilon b_i + \sum \epsilon_i b_i^2| \leq \epsilon^2 + 2\epsilon||b|| + |\sum \epsilon_i b_i^2|.$$

By Höffding inequality, $\sum \epsilon_i b_i^2$ has orlicz norm for Ψ_2 bounded by a multiple of $(\sum, b_i^4)^{1/2} \leq \eta_n(\sum_i^n b_i^2)^{1/2}$. Apply Lemma 6.7 to the third term on the right and substitute $\sup A_n$ for sup over B_n.

$$E_\epsilon ||\sum \epsilon_i a_i^2|| \leq \epsilon^2 + 2\epsilon||\sum a_i^2||_{A_n}^{1/2} + \sqrt{1 + \log|B_n|}\eta_n||\sum a_i^2||_{A_n}^{1/2}.$$

The size of the ϵ-net can be chosen

$$|B_n| \le N^2(\epsilon/2, \mathcal{F}, dn)$$

by entropy condition d, this variable is bounded in probability.

Conclude for some constant, K

$$P(||\sum \epsilon_i a_i^2||_{A_n} > t) \le P^*(B_n| > M) + K/t[\epsilon^2 + (\epsilon + \eta_n \sqrt{\log M}) x E||\sum a_i^2||_{A_n}^{1/2}].$$

For M, t sufficiently large, the RHS is smaller than $1 - v$, for the constant v, in Proposition 6.5.3. More precisely, this can be achieved for M such that

$$P^*(B_n| > M) \le (1 - v_1)/2$$

and $t = (1 - v_1)/2$ and t is equal to $(1 - v_1)$ times the numerator of second term. Then t is bigger than v-quantile of $|| \sum \epsilon_i a_i^2||_{A_n}$. Thus, Proposition 6.5.3 gives

$$E||\sum \epsilon_i a_i^2||_{A_n} \lesssim E|| \max a_i^2||_{A_n} + t \lesssim \eta_n^2 + \epsilon^2 + (\epsilon + \eta_n \sqrt{\log M})(E||\sum a_i^2||_{A_n})^{(1/2)}.$$

Now (c) gives $|| \sum E a_i^2 || A_n \to 0$. Combining this with Lemma 6.5.1, we get

$$E||\sum a_i^2||_{A_n} \le E||\sum \epsilon_i a_i^2||_{A_n} + 0(1) \le \delta + \delta(E||\sum a_i^2||_{A_n})$$

for $\delta > max(\epsilon^2, \epsilon)$ and sufficiently large n.

Now for $c \ge 0$, $c \le \delta + \delta\sqrt{c}$ implies

$$c \le (\delta + \sqrt{\delta^2 + 4\delta})^2.$$

Applying this to $c = E|| \sum a_i^2||_{A_n}$ to conclude $E|| \sum a_i^2||_{A_n} \to 0$ as $n \to \infty$.

Example 1: One can look for $X_1 \ldots X_n$ i.i.d. random variables

$$Z_{ni}(f) = n^{(-1/2)} f(X_i)$$

with f measurable functions on the sample space. From the above theorem, we get that $\mathcal{F}$ is Donsker if it is measurable and totally bounded in $L_2(P)$, possesses a square integrable envelop and satisfies with P_n equal to distribution of $X_1 \ldots X_n$

$$\int_0^{\delta_n} \sqrt{N(\epsilon, \mathcal{F}, L_2(P_n))} d\epsilon \xrightarrow{P} 0.$$

If $\mathcal{F}$ satisfies the uniform entropy condition in section 6.5, then it satisfies the above condition. Thus, Theorem 6.10 is the consequence of the above theorem.

Example 2: The Lindberg condition on norms is not necessary for the CLT. In combination with truncation the preceding theorem applies to more general processes. Choose stochastic processes $Z_1 \ldots Z_n$, m_n with

$$\sum_1^{m_n} P(||Z_{ni}||_{\mathcal{F}} > \eta) \to 0 \text{ for } \eta > 0.$$

Then truncated processes $Z_{ni}, \eta(f) = Z_{ni}(f)1\{||Z_{ni}|| \leq \eta\}$ satisfy

$$\sum_1^{m_n} Z_{ni} - \sum_1^{m_n} Z_{ni}, \eta \xrightarrow{P} 0 \text{ in } l^{\infty}(\mathcal{F}).$$

Since this is true for every $\eta > 0$, it is also true for $\eta_n \downarrow 0$ sufficiently slowly. The processes Z_{n,i,η_n} satisfy Lindberg condition. If they or their centered version satisfy other conditions of Theorem 6.5.1, then the sequence

$$\sum_1^{m_n} (Z_{ni} - EZ_{ni,\eta_n})$$

converges weakly in $l^{\infty}(\mathcal{F})$. The random semi-metrics d_n decrease by truncation. Thus, one can get the result for truncated processes under weaker conditions.

We now define measure-like processes. In the previous theorem, consider index set $\mathcal{F}$ as a set of measurable functions $f : x \to \mathbb{R}$ on a measurable space (x, a) and the distance be in $L_2(Q)$. Q finite measure: assume

$$(1) \quad \int_0^{\infty} \sup_Q \sqrt{\log N(\epsilon ||F||_{Q,2} \mathcal{F} L_2(Q))} d\epsilon < \infty.$$

Then the preceding theorem yields a central limit theorem for processes with increments that are bounded by random L_2-metric on $\mathcal{F}$. We call Z_{ni} measure-like with respect to random measures $\mu_{ni} : f$

$$(Z_{ni}(f) - Z_{ni}(g)) \leq^2 \int (f-g)^2 d\mu_{ni} f, g, \epsilon \mathcal{F}.$$

For measure-like processes d_n is bounded by $L_2(\sum \mu_{ni})$-semi-metric and entropy condition there can be related to uniform entropy condition.

Lemma 6.5.3: Let $\mathcal{F}$ be a class of measurable function F. Let $Z_1, \ldots Z_n$, m_n be measurable processes indexed by $\mathcal{F}$. If $\mathcal{F}$ satisfies uniform entropy condition (1) for a set Q that contain measures μ_{ni} and $\sum_{1}^{m_n} \int F^2 d\mu_{ni} = 0_p^*(1)$ then entropy condition d) of Theorem 6.5.1 is satisfied.

Proof: Let $\mu_n = \sum_{i=1}^{m_n} \mu_{ni}$. Since d_n is bounded by $L_2(\mu_n)$-semi-metric we get

$$(*) \quad \int_0^{\delta_n} \sqrt{\log N(\epsilon, \mathcal{F}, d_n)} d\epsilon \leq \int_0^{\delta/||F||_{\mu n}} \sqrt{\log N(\epsilon||F||_{\mu n}, \mathcal{F}, L_2(\mu_n))} d\epsilon . ||F||_{\mu n}$$

on the set where $||F||_{\mu_n}^2 < \infty$. Denote by

$$J(\delta) = \int_0^{\delta} \sup_Q \sqrt{\log N(\epsilon||F||_{Q,2,} \mathcal{F}, L_2(Q)} d\epsilon$$

on the set where $||F||_{\mu_n} > \eta$ the RHS of the equation $(*)$ is bounded by $J(\delta_n/\eta) O_P(1)$ which converges to zero for $\eta > 0$. On the set where $||F||_{\mu_n} \leq \eta$, we have the bound $J(\infty)\eta$ as η is arbitrary we get the result.

Example 1: Suppose we have independent processes $Z_1, \ldots Z_n$, m_n are measure line. Let $\mathcal{F}$ satisfy uniform entropy condition. Assume for some probability measure P with $\int F^2 dP^* < \infty$

$$E^* \sum_{1}^{m_n} \int F_{ni}^2 1\{\int F^2 d\mu_{ni} > \eta\} \to 0 \text{ for } n \to 0$$

and

$$\sup_{||f-g||_{p,2} < \delta_n} E^* \sum_{1}^{m_n} \int (f-g)^2 d\mu_{ni} \to 0 \text{ as } \delta_n \downarrow 0$$

and

$$\int F^2 d\mu_{ni} = 0_p^*(1).$$

then we get that condition of Theorem 6.5.1 (d) is satisfied and other conditions imply the other conditions as

$$||Z_{ni}||_{\mathcal{F}} \leq Z_{ni}(f)1 + 4 \int F^n d\mu_{ni} \text{ for any } f.$$

This gives $\sum_{1}^{m_n}(Z_{ni}-EZ_{ni})$ converges in $l^{\infty}(F)$ if covariance functions converge point-wise and measureability conditions as satisfied.

Example 2: Consider in the above example $Z_{ni} = c_{ni}\delta_{x_{ni}}$ where for each n, (c_{ni} constant) $X_1, \ldots X_n$, m_n are independent random variables is a measurable space (x, a) with induced measures P_{ni} with $\int fdP_{ni}$ exist for each element f of a class of measurable functions $f : x \to \mathbb{R}$. We can consider weighted empirical process

$$G_n(f) = \sum_{i=1}^{m_n} c_{ni}(f(X_{ni}) - \int fdP_{ni}).$$

Suppose $\max_{1 \leq i \leq m_n} |c_{ni}| \to 0$

$$\sum_{1}^{m_n} c_{ni}P_{ni} \leq P$$

for P a probability measure P with $E_p^* F^2 < \infty$. Note that Z_{ii} are measure-like with $\mu_{ni} = c_{ni}^2\delta_{x_{ni}}$. Then we get under measurability condition G_n converges weakly to a Gaussian process on $l^{\infty}(\mathcal{F})$. In addition, we can prove that the limiting process has uniformly continuous sample paths with respect to $L_2(P)$-semi-metric.

Bibliography

[1] Aldous D. Stopping times and tightness. Ann Probab. 1978;6:335–340.
[2] Billingsley P. *Convergence of probability measures.* New York: Wiley; 1999.
[3] Billingsley P. The invariance principle for dependent random variables. Trans Amer Math Soc. 1956;83:250–268.
[4] Chari R, Mandrekar V. On asymptotic theory in censored data analysis without independent censoring, RM-431. Statistics and Probability. MSU; 1983.
[5] Donsker M. An invariance principle for certain probability limit theorems. Mem Am Math Soc. 1951;6:12 p.
[6] Dudley RM. *Uniform central limit theorems.* Cambridge: Cambridge University Press; 1999.
[7] Dudley RM, Phillips W. Invariance principles for sums of Banach space valued random elements and empirical processes. Z Wahrsch Verw Gebiete 1983;62:509–552.
[8] Durrett R. *Probability: theory and examples.* 3rd ed. Belmont, CA: Thomson (Brooks/Cole); 2005.
[9] Durrett R, Resnick S. Functional limit theorems for dependent random variables. Ann Probab 6:829–846.
[10] Dvoretsky A. Asymptotic normality for sums of dependent random variables. In Proc 6th Berkeley Symp Vol II; 1972. p. 513–535.
[11] Dynkin EB, Mandelbaum A. Symmetric statistics, Poisson point processes and multiple Wiener integrals. Ann Stat 1983;11:739–745.
[12] Gill R. Large sample behavior of product limit estimator on the whole line. Ann Stat. 1983;11:49–58.
[13] Gordin MI. The central limit theorem for stationary processes. Soviet Math Dokl. 1969;10: 1174–1176.
[14] Gross L. *Harmonic analysis on Hilbert space.* Mem Am Math Soc. 1963;46:ii+62.
[15] Jacod J, Shiryayev AN. *Limit theorems for stochastic processes.* Berlin: Springer-Verlag; 1987.
[16] Lenglart E. Inequality de semimartingale. Ann Sci Univ Clermont-Ferrand II Math. 1981;19: 77–80.
[17] Liptser RS, Shiryayev A. Necessary and sufficient conditions in the functional central limit theorem for semimartingales. Theory Prob Appl 1981;26:132–137.
[18] Mandrekar V. Central limit problem and invariance principle in Banach Space Seminaire des Probabilities XVII, *Lecture Notes 986.* Springer-Verlag; 1983.
[19] Mandrekar V. Mathematical work of Norbert Wiener. Notices Am Math Soc. 1995;42:364–369.
[20] Mandrekar V, Rao BV. On a limit theorem and invariance principle for symmetric statistics Prob Math Stat 1989;10:271–276.
[21] Mandrekar V, Thelen B. *Joint weak convergence on the whole line in the truncation model,* In Proc R C Base Symposium on Probability, Statistics, and Design of Experiments. New Delhi: Wiley (Eastern); 1990. p. 495–515.
[22] Prokhorov Yu V. Convergence of random processes and limit theorems in probability. Theory Prob Appl. 1956;1:157–214.
[23] Skorokhod AV. *Studies in the theory random processes.* Reading, MA: Addison-Wesley; 1965.
[24] Susarla V, Van Ryzin J. *Large sample theory for survival curve estimators under variable censoring. Optimization methods in statistics.* New York: Academic Press; 1976. p. 475–508.
[25] Van der Vaart A, Wellner JA. *Weak convergence and empirical processes (with applications to statistics).* New York: Springer; 2000.
[26] Woodroofe M. Estimating a distribution with truncated data. Ann Stat 1985;13:163–177.
[27] Sazanov, VV. *On characteristics functionals Theory.* Prob Appl 1958;3:201–205.

www.ingramcontent.com/pod-product-compliance
Lightning Source LLC
Chambersburg PA
CBHW082035230326
41598CB00081B/6514